Environmental planning
perception and behavior

Environmental Planning

Perception and Behavior

Thomas F. Saarinen
The University of Arizona

Houghton Mifflin Company Boston
Atlanta Dallas Geneva, Illinois Hopewell, New Jersey Palo Alto London

Cover artwork adapted from F. W. Boal, "Territoriality on the Shankill-Falls Divide, Belfast," *Irish Geography*, Vol. 6, No. 1 (1969).

Printed in the U.S.A.

Library of Congress Catalog Card Number 75-19533

ISBN 0-395-20618-9

To my parents
who taught me to see things
from more than one perspective.

Contents

Preface

The title of the book, *Environmental Planning: Perception and Behavior,* was selected to reflect the general scope of this rapidly growing, interdisciplinary field of study that seeks to combine the insights of social and behavioral sciences with the skills of the design and planning disciplines. A conviction shared by researchers in the field is that to understand environmental decision-making, knowledge of peoples' environmental perceptions and behavior is essential. The conviction stems from the observation that decisions to modify or change the environment are based not so much on the environment as it is but rather on the environment as it is perceived or conceived by the decision-maker.

The book is intended for use in a wide range of courses in the social and behavioral sciences. Because it draws on several disciplines, no special knowledge of any one discipline is requisite on the part of the reader. *Environmental Planning* is appropriate for a number of introductory courses, including introductory geography, environmental planning, environmental studies, and human ecology. It may be of particular interest to teachers and students of environmental perception, spatial behavior, and human engineering, to name just a few of the related fields.

It has been a fascinating personal experience for me to watch the mushrooming growth of this field, as I have watched it emerge and evolve from the time I first entered graduate school at the University of Chicago—with my intention even then of combining the study of psychology with the study of geography. I contrast my own introduction to the field in the early 1960s with that of the students in my classes today. In the 1960s there were only a few, tantalizing ideas floating around. Isolated individuals were beginning to venture beyond the boundaries of their own specializations to explore this new interdisciplinary territory. Now some fifteen years later a veritable flood of printed material on environment and behavior pours over my desk, including books, journals, articles, descriptions of upcoming conferences, directories, and newsletters. The materials clearly indicate a proliferation of approaches as each of the earliest staked-out research areas is divided and subdivided into myriad new territories. To the curious novice exploring the field, this diversity must seem exciting but also bewildering.

This book is written to provide the novice with a concise introduction to the field in the form of a broad survey of the first major studies and experiments. Hopefully, the undergraduate or beginning graduate student will gain the perspective necessary to set off toward the field's frontiers and explore the more recent works of a developing subfield.

Other books on this subject have encompassed only portions of the entire field, focusing on only one particular scale or one professional perspective. This book is organized by scale so that it leads to more comprehensive coverage of the various subfields and different disciplinary approaches. It should prove useful not only for novices but also for those working in specialized subfields who wish to learn of the developments and discoveries in related but sometimes remote areas. A difficult task confronting the reviewer of this particular combination of social science and design research is that of reconciling the contrasting points of view of these two groups of professionals. Books by social scientists on environmental planning or perception tend to concentrate on theoretical issues while books by designers seek practical applications. I have made a deliberate effort to select studies that are oriented both to the theoretical issues involved and to the development of guidelines that could be useful in deciding what to do with the practical problems planners confront every day.

To acknowledge fully all the help I have received would involve retracing all the personal and professional contacts of my career. I will not attempt to do so although I am duly grateful to all of them. I am grateful to the University of Arizona because without my recent sabbatical leave from the University, the book would not have been possible. Also, I would like to thank especially Mrs. Marian Clark and my wife Caryl who handled the bulk of the typing through many stages.

Several people reviewed the manuscript at various times in its development, and the book is surely better for their helpful suggestions: Edward Ostrander, Cornell University; Richard Parish, Cuyahoga Community College; Ira Firestone, Wayne State University; Roger M. Downs, Pennsylvania State University; Kenneth H. Craik, University of California, Berkeley; and John T. Matson, State University of New York at Cortland.

Thomas F. Saarinen

One

Introduction

"The long-range question is not so much what sort of environment we want, but what sort of man we want."
Robert Sommer

The greatest challenge facing us today is how to live in harmony with our environment on spaceship earth. Whether we want to or not, we are creating a new world, and in so doing, we are creating a new person. To house the world's projected population for the year 2000 will require building in one generation more structures than we have built in the whole of human history.[1] This means that the artificial portion of the environment (the built environment) will contain more buildings, neighborhoods, cities, and great urban regions than it ever had before. Not only will the built environment be more extensive, but it will be changed in such a way that, of necessity, people will have to adjust to it.

The natural environment outside artificial areas will be stressed increasingly as we clear the earth, drain it, replant it, deforest it, reforest it, spray it, dig it, dump on it, drill it, level it, irrigate it, and otherwise alter it to obtain the materials necessary to feed, clothe, and shelter adequately the multiplying millions of humans. With greater pressure on the natural environment, we will have to make sensitive adjustments to avoid augmenting the largely negative and inadvertent side effects of our actions, which are evident already in the forms of increased air pollution, water pollution, and land pollution. Greater population numbers and densities,

increasing interdependence among world areas and peoples, more rapid transportation and communication systems, all will cause vast changes in the type and amount of contact we have with that portion of our environment made up of other people (social environment). Whether considering the built environment, the natural environment, or the social environment, it seems clear that fundamental changes are likely to occur in people-environment interactions. To create an improved world, it is essential that this people-environment relationship be examined directly and understood so that we can make wise decisions in planning future alterations of ourselves and our environment.

To create an environment that will aid in the evolution of a better society is a worthy goal, but unfortunately we do not know how to do so. We know next to nothing about people-environment interactions, and we are only just beginning to learn how to investigate the problems. Before we can plan, we must come to grips with a whole range of problems intrinsic to people-environment relations. For example, how can we measure the environment and its effects on us? How can we measure our perception of the environment, which is often quite different from the objective environment? How can we measure and account for the effects of individual and cultural differences on environmental perception? How can we determine what our goals and needs are for the environment and what the optimum or limiting factors are for environmental comfort or stress? How do we collect and mentally organize environmental information? In the following chapters, we will discuss some of the early attempts to study such problems; and in so doing, we will survey this vast, rapidly expanding, multidisciplinary area of research labeled by various researchers as environmental psychology, environmental perception, environmental behavior, environics, human ecology, environmental biomedicine, spatial behavior, man-environment relations, humanics, sociophysical design, ecological psychology, environmental physiology, human engineering, environmental design, environmental science, ergonomics, urban environmentalism, sociological psychology, behavioral geography, psychological ecology, psychogeography, and man-environment systems. The field referred to by these terms varies from narrowly defined, highly specialized subfields to all-inclusive, general frameworks. But all share certain similar characteristics that will become apparent in the course of our discussion.

Canvassing the many fields of inquiry included in this survey reveals that the result is unlikely to be an in-depth study that evenly weighs all the types of studies or assigns proper significance to them. Rather, it reflects my own reading, and since I am a geographer, the survey could be considered a personal geographic perspective of the field of environmental perception and behavior. I have personally conducted research on such topics as perception of the drought hazard on the Great Plains, the image of the Chicago Loop, perception of environmental quality in Tucson, Arizo-

na, and student views of the world. I have collaborated with researchers on closely related topics and conversed with others on more distant ones. But for many of the topics included here, my only sources of information were the written reports of the researchers.

This survey is clearly an exercise in "foolrushery," as defined by J. K. Wright, who coined the term to mean "what fools do when they rush in where angels fear to tread."[2] Wright justified foolrushery as valuable for promoting the cause of interdisciplinary cross-fertilization, adding that "interdisciplinary cross-fertilization is as indispensable to the balance of scholarship as the bees are to the balance of nature."[3] In spite of the obvious limitations of a survey like this one, I am still hopeful that it will help interdisciplinary communication.

Organization of the book

The chapters in this book are organized according to scale. Starting at the smallest scale, the chapters in turn discuss studies on our relationship to progressively larger units of our environment. We begin with personal space and room geography and go on to architectural space, neighborhoods, cities, larger conceptual regions, countries, and the world. At each scale the discussion focuses on some of the most representative of the studies completed to date. These studies were selected to illustrate clearly how certain aspects of the interaction between us and the environment are thought to operate in everyday life. Some of the earliest or best known of the early studies were selected to provide a perspective on the direction of research development at each scale and to demonstrate the methods used. Separation of the studies by scale highlights the distinct contributions of different disciplines, for each tends to focus on a scale that reflects its predominant concern and perspective. Environmental behavior at each scale considered seems to operate as a separate system, but striking similarities pervade the entire gamut from our conception of personal space to our conception of the world.

None of the results of the studies are considered final or definite, to be stated as proven facts. Rather, they illustrate the types of approaches that could potentially yield fruitful results, and they are worth testing further for these possible future benefits. Even if they do not turn out as expected, they should help advance our current understanding. In many cases the studies are presented in broad-stroke outline, omitting qualifications. This is not meant to deceive the reader into thinking that the results are conclusive but rather to highlight the key points of the research.

The smaller units of environment, particularly the social environment, might be considered the domain of the psychologists. This is evident in Chapter 2, on personal space and room geography, where a predominant concern is exploring some of the psychological dimensions of human use of

space. Much of the research at this scale is conducted by psychologists who are concerned primarily with the portion of the environment made up of other people and with the way an individual's behavior may be structured by spatial positions of other people and by physical arrangements of furniture within a room. When other disciplines work at this scale, they bring their own perspectives to the research. Thus the anthropologist is concerned with the cross-cultural differences in personal space; the sociologist, with the use of environmental props to help in playing social roles; and the industrial designers and psychological engineers, with measuring the physical dimensions of people to improve the design of machines, rooms, and other small spaces. Chapter 2 illustrates these differences in disciplinary perspectives in discussing topics such as proxemics, human factors in industrial design, invasion of personal space, territoriality within rooms, and the effects of seating positions on the communication process.

Psychologists are also active at the scale of architectural space, discussed in Chapter 3. But, as might be expected, much impetus for research has come from architects and other members of the design professions who are faced with the practical task of designing better buildings. The designers are concerned with the problem of translating the implications of behavioral science research into a tangible physical form. It is therefore our relationships to various dimensions of the built environment that are explored. Some aspects of hospital design are used to illustrate the research linking human needs with architectural design. The varied perspectives of doctors, nurses, patients, and administrators become apparent in discussing optimal design for nursing units, function as a basis for psychiatric ward design, behavioral mapping, territoriality in a psychiatric ward, and medical center planning in the context of the surrounding community.

At the scale of the neighborhood or the small town, Chapter 4, a new disciplinary group, the sociologists, become dominant, and they bring to the discussion their concern with social groups. The distinct differences in the images of environment held by various subcultural groups become a focus for research. The environment considered at this scale is mainly the built environment. It is interesting to contrast the focus of the sociologists on the social group with two other disciplinary perspectives on the small town. The ecological psychologists try to measure the psychological atmosphere of the town as it impinges on the individual, while the geographer's interest centers on the location and arrangements of landscape features. The major topics covered in Chapter 4 are the relationship between site planning and social behavior, small town atmosphere, highway strip developments, the work of ecological psychologists, the nature of the neighborhood attachment of urban slum dwellers, and cognitive maps of neighborhoods.

Fascination with the problem of how people perceive and react to the city, the largest and most complex of human artifacts, has attracted interest

from many fields. Sociologists, anthropologists, geographers, psychologists, novelists, lawyers, historians, psychiatrists, architects, public health officials, and many others have all made contributions. It is at the city scale, discussed in Chapter 5, that one finds the greatest amount of overlap among the disciplines researching environmental perception and behavior. City planners are, of course, represented with their unique perspective on the design and physical structure of entire urban areas. But in addition, other researchers represent disciplines that are generally more active at smaller or larger scales, such as psychologists, who are trying to measure the psychological atmosphere of cities, and sociologists, who are stressing the varied perspective and roles of social groups and the social meanings of city areas. Geographers and anthropologists, who are generally more active in larger units of the environment, also extend their range in scale to include an interest in the city. These varied perspectives on the city are evident in Chapter 5 in the discussion of the image of the city, mental maps, urban symbolism, subjective distance in cities, urban activity systems, and components of the urban atmosphere such as noise and crowding.

From the topics considered in Chapter 5 on the city, we shift from direct perception of the immediate surroundings, still largely the built environment, to people's conceptions or ideas of the environment. Thus we see discussion of images, mental maps, and urban symbolism. As larger units of the environment are considered in succeeding chapters, we will concentrate more on conceptions of the environment, mental images, ideas, attitudes, and so on. This shift in emphasis is necessary because the larger the size of the units, the less possible it is to see the entire unit at once or to know all its aspects intimately.

In Chapter 6, where we consider larger conceptual units of the environment, we find different disciplines conducting the research. This scale between the city and the nation is the special realm of geographers, who are also active at the scale of the nation and the world. Their ability to define significant regions according to such attributes as climate, river basins, political beliefs, cultural backgrounds, and agricultural practices stems from their focus on the location and spatial arrangement of phenomena. Anthropologists are also active at the regional scale that coincides with the size of most culture areas. Historically, at this scale sociologists have studied the social group differences in meanings attached to various recreational activities in national parks, forests, and wilderness areas. At this scale emphasis is on the natural environment: perception of natural hazards, recreation and wilderness studies, cognitive anthropology, imagined areas of the past, environmental personality, and research on human perception of weather and climate.

The shift from measurement of direct perception of immediate surroundings to people's conceptions of the environment is virtually complete at the scale of the nation, discussed in Chapter 7. For example, when landscapes

are considered at the scale of the nation, the main concern is no longer with the physical features or people's reactions to them. Instead attention is devoted to the effect of national attitudes and preferences on the development of national landscapes. The focus on more abstract images, ideas, or conceptions is reflected in some of the topics of the chapter: landscape preferences, the effect of landscape preference on public policy, place preferences, regional consciousness, and national character.

The high level of abstraction intensifies in Chapter 8, which deals with the international scale. Here political scientists study perception of the entire world or with very large segments of it that include more than one nation. International relations, with the illusive concept of power as a major factor, complete the gradual shift from reality to image, for at this scale a symbolic environment is not always objectively measurable. The topics in Chapter 8 are Western attitudes toward nature, world views in the past and present, images of the world and various world areas, and cross-cultural study of attitudes.

The remainder of Chapter 1 presents some basic definitions and characteristics of the field along with a schema illustrating some theories and concepts that underlie current research in environmental perception and behavior. Chapter 9 contains some very general conclusions drawn from a broad comparison of the studies of environmental perception and behavior from the scale of personal space to the world.

Some basic definitions

The *American Heritage Dictionary* defines *environment* as "the combination of external or extrinsic physical conditions that affect and influence the growth and development of organisms." However, the organism and the environment represent interacting systems, and it is not always easy to draw the line between them. To illustrate, consider some questions raised by Bates in his discussion of environment in the *International Encyclopedia of the Social Sciences.* When does an apple being eaten cease to be part of the environment and become a part of the individual? Is an internal parasite a part of the environment? Should culture be considered a part of the individual or the environment? An ecological approach is considered to be the most useful in this book. Each of the different scales of people-environment interaction is thought of as a separate human ecosystem that is a total system including the human members as well as their surroundings. This means that a change at any point would have reverberations throughout the system. At different points in the following discussion there is an implicit, if not explicit, focus on different components of the environment, such as, the *social environment* composed of other people, the *built environment* consisting of the physical structures we erect; and the *natural environment* including weather, climate, and all the other physical

processes of the earth. For the purposes of studying human environmental perception and behavior, it is useful to focus on the *functional environment*—that is, the portion most pertinent to the people being studied. Sonnenfeld provides a thoughtful discussion of this behavioral environment and defines it in terms of a nested set of environments (Figure 1.1).[4] The whole environment that is external to human beings, the entire world, is the objective, *geographical environment.* The awareness may be derived from learning and experience or from physical sensitivity to environmental stimuli. Thus at this level a portion of the environment is symbolic rather than objectively measurable. The least-inclusive level is the *behavioral environment,* the portion that elicits a behavioral response or toward which behavior is directed. In distinguishing the behavioral environment, one's thinking is directed toward a consideration of the behaviorally significant versus the behaviorally insignificant elements of the environment. This selection problem is a major concern of studies of environmental perception and behavior.

Perception is an extremely complex concept. The new *International Encyclopedia of the Social Sciences* devotes more than fifty pages to discussing its meaning, but the studies reviewed here are mainly concerned with one of the 10 subfields concerned with social perception. Even this narrower set of meanings is sufficiently complex to require eight pages of discussion. In the simplest terms *social perception* is generally concerned with the effects of social and cultural factors on our cognitive structuring of our physical and social environment. Perception then depends on more than the stimulus present and the capabilities of the sense organs. It also varies with the individual's past experiences and present "set" or attitude acting through values, needs, memories, moods, social circumstances, and expectations. The major problem in studying people's perception is that of measurement, since people often have difficulty articulating the conscious or unconscious feelings, attitudes, or ideas associated with perception. In many cases perception must be inferred, from behavior or from other indirect sources.

Behavior, as used in the term "behavioral sciences," is a new term, which came into currency in the 1950s largely through an administrative action of the Ford Foundation establishing a behavioral science program. The purpose of the program was to increase knowledge of human behavior through problem-oriented scientific research that could be utilized in human affairs. *Environmental behavior* would include not only the overt response of the individual or group to environmental factors but also subjective behavior, such as attitudes, beliefs, expectations, motivations, and aspirations.

Planning may be considered the conscious organization of human activity to serve human needs. Better planning can be accomplished by greater integration of the separate components at each scale into a broader, more coherent framework. To be effective, planning must consider not only the

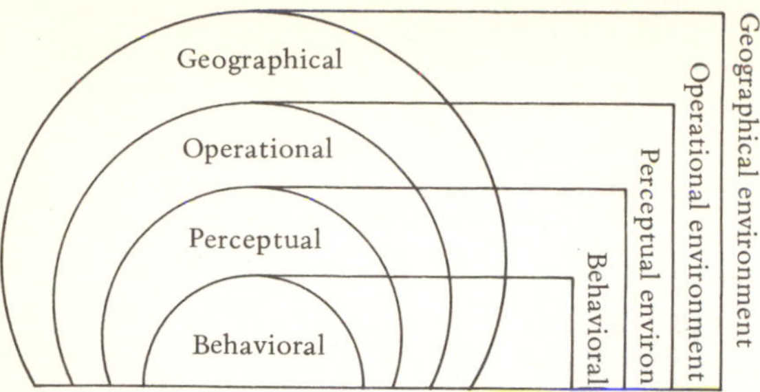

Fig. 1.1 Nested set of environments. (Joseph Sonnenfeld, "Geography, Perception, and the Behavioral Environment," Paper at AAAS Dallas Meeting, December 27, 1968. Adapted by permission.)

physical environment but the way people perceive and utilize each segment of the environment. Changes in the social and physical arrangements may range from small adjustments to redesign of an entire society. An ultimate goal would be a unified organization of social life for all people. Whatever the scale, such planning requires great stress on the evaluation of results and on use of objective measures of success or failure.

Characteristics of the field

Some of the major characteristics of the field of study called "Environmental Planning: Perception and Behavior" are: the recency of its development, the lack of well-developed methodology, its interdisciplinary nature, and the prominence of planning or concern for current environmental issues.

The development of studies on environmental perception and behavior is so recent that no real body of theory has developed. Nor has a single name been agreed upon for the field. Although the roots of ideas and concepts can be traced back several decades, the earliest suggestions for their use usually appeared as incidental aspects of other studies. It is only in the last decade that environmental perception and behavior have become the main focus of research and only in the last five years that the bulk of the studies has appeared. Some of the milestones in the development of the field are: the first appearance of a directory of researchers in 1965,[5] the inaugural issue in 1969 of *Environment and Behavior,* an interdisciplinary journal devoted exclusively to reporting scholarly work on the reciprocal relationships between the physical environment and behavior, and the first comprehensive attempt to define the boundaries of the field in a book of readings published in 1970.[6] It is not surprising then that no real body of theory has developed thus far. But the rapid accumulation of empirical research and the growing number of researchers attest to its viability.

The lack of a well-developed methodology in the studies of environmental perception and behavior stems from recency of its development as well as the type of behavior investigated. Instead of the traditional experimental approach to the study of behavior that emphasizes controlled conditions and limited aspects of experience, these studies try to abstract or generalize from total behavior in real-life situations. Problems in selection and interpretation are inevitable under such circumstances. Barker and Wright in their study of the 119 children in Midwest, Kansas, estimated that their subjects engaged in more than 36 million behavioral episodes in the course of a year.[7] Clearly, it is impractical if not impossible to record all behavior, so some selection must be made. Once such a selection of types of behavior has been made, however, the problem of interpretation still remains. It is difficult to explain behavior in terms of cause and effect because there are many missing links in the chain of events from environment to the individual and back. What is involved is a choice between roughly outlining the significant factors in real human decisions related to environment with the hope of developing theory later or precisely measuring limited and isolated aspects of behavior in artificial environments to build up gradually a more rigorous methodology and theory that can then be applied to the real world. The researchers reviewed in this book have generally opted for the former alternative because they are concerned with current environmental problems and because they have some doubt that laboratory findings can be transferred readily to the more complicated circumstances of the real world.

Current environmental problems and the search for planning applications make up another characteristic of the field. Whether the study centers on urban neighborhoods, residential design, resource management, or international behavior, there tends to be a strong emphasis on providing information useful for public policy decisions. A key idea is to design with human beings as the measure, in hopes of solving or at least ameliorating many pressing social problems. Every day decisions are made with respect to these problems regardless of a lack of theory on people-environment interactions. It seems likely that the quality of these decisions can be improved by the provision of better information on how people perceive and react to their environment.

A final characteristic to be noted here is the interdisciplinary nature of the field. In the search for effective methods of measuring environmental perception and behavior, researchers have been motivated to go beyond the conventional boundaries of their disciplines. In addition, the problems have been recognized and solutions sought by scholars from many different disciplines. Much of the work represents a fusion of ideas from many fields that creates a stimulating research atmosphere but causes problems as well, such as, trying to reconcile fundamental differences in disciplinary philosophies, vocabularies, and research directions. However, the free flow of methods, concepts, and measuring techniques across disciplinary

lines suggests that the focus on people-environment systems may prove a unifying factor for the social and behavioral sciences. It should also link them more closely to earth sciences and to the planning and design professions. There is evidence that many researchers are attempting to bridge the gulfs separating their disciplines from others by placing their research within a broader social and behavioral science framework. Such an attempt is contained in following the conceptual schema developed by Roger Downs.[8]

A conceptual schema

The Downs schema incorporates a number of the basic theories and concepts that underlie current research in environmental perception and place it in a broad social science context. Thus people are viewed as decision-makers, their behavior is considered to be some function of their image of the real world, and they are regarded as complex information-processing systems (Figure 1.2).

The basic process in interaction in the schema is as follows: The real world is taken as the starting point, and it is represented as a source of information. The information content enters the individual through a system of perceptual receptors, and the precise meaning of the information is determined by an interaction between the individual's value system and their image of the real world. The meaning of the information is then incorporated into the image. On the basis of this information, the individual may require to adjust himself with respect to the real world. This requirement is expressed as a decision which can, of course, be one that involves no overt reaction. The links from the concept of a decision are two-fold (although these could be amalgamated). The first link is a recycling process, called search, whereby

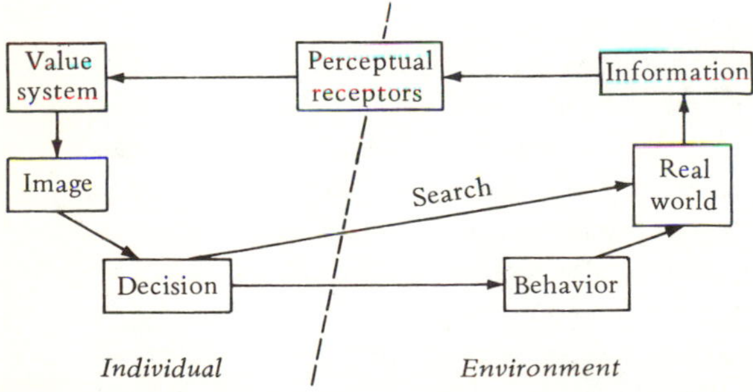

Fig. 1.2 A conceptual schema for research into geographic space perception. (Roger M. Downs, "Geographic Space Perception," *Progress in Geography*, 2, Edward Arnold, publishers, Ltd., London, 1970, p. 85. Adapted by permission.)

the individual decides that sufficient information has been acquired, or some time-cost limitation acts as a constraint to further search. A decision is then made which may be expressed as a pattern of behavior which will in turn affect the real world. Since the real world undergoes a change, fresh information may result, and the whole process can continue. The schema therefore allows the space perception process to occur in a temporal as well as a spatial context.[9]

Downs suggests viewing the factors that cause people to see the same segment of the real world differently as a set of filters. The constant input of information is then seen as it is screened by the physiological filters of our sense receptors and such important psychological filters as language, social class, personal values, value and need, culture, and some form of gestalt or pattern-seeking function.

The communication process

The power of these psychological filters to alter or select from the incoming information may be illustrated by the example of language, the primary communication system. In the process of adjusting to the environment, each culture selects from an infinite array of possibilities a certain set of categories to describe and explain what is there. These categories become part of their system of communication and thus structure for succeeding generations which aspects of the environment are attended to. A familiar example is the contrast between English, which has only one word for snow, and the Eskimo language, which includes a great variety of words that indicate different qualities, such as falling snow, snow on the ground, snow packed hard like ice, slushy snow, wind-driven flying snow, and so on.[10] Pitts provides some interesting examples of the degree of differentiation of certain natural phenomena provided by the Japanese words for types of paddy fields, winds, rain, and snow.[11] Table 1.1 indicates the sharp observation, attention to detail, and subtle differences in humidity, temperature, and speed that can be expressed by various words for rain in the Japanese language. In the examples provided, the elaborate vocabulary developed indicates the importance of the phenomena to the social group considered. In different languages there are great differences, not only in vocabulary but in grammatical structure as well, that may totally change the usual way of viewing aspects of experience such as time, space, or relationships to nature or other people. But that is only the beginning of the problem, for there are problems of communication stemming from language not only between people speaking different languages but also between speakers of the same language.

Within any linguistic group the same term may come to have a variety of meanings for different individuals and groups. Meredith Burrill's work at the Board on Geographic Names provides some fascinating examples.[12] He

Table 1.1 Japanese Words for Types of Rain

yūdachi	Squall, usually with lightning
shigure	Cold, drizzling rain in autumn
samidare	Intermittent, almost continual rain in and around May
tsuyu	Same; another name for what the Chinese call "plum rains"
niwakaame	Sudden rainfall
itosame	Rain that comes down like fine threads
nukaame	Drizzle as fine and soft as rice bran
kirisame	Misty rain
doshaburi	Downpour that comes down quite hard
ōame	A big rain
kosame	A slight rain
harusame	Spring rain around March or April
murasame	A steady rain which doesn't last long, but longer than a squall
hisame	Icy rain in winter

SOURCE: Forrest Pitts, "The Japanese Landscape as Mood and Intensity," unpublished manuscript, 1960, p. 9.

warns that "one can easily fall into the trap of regarding categories as entities whose reality is independent of human invention."[13] The connotations of geographic terms—for example, summit, valley, plain, desert, gorge, swamp, archipelago, and creek—vary depending on what examples we have seen and what our parents or other authorities may have told us or shown us early in life. Thereafter a great reluctance to change the meaning is apparent, and a surprising amount of anger and frustration is aroused when concepts are challenged. Burrill illustrates the difficulty of understanding other people's categories by noting that,

> . . .the Atlantic coast swamp with all those attributes is a sort of complex, the nature of which is clearly known to the local people who are parties to a tacit understanding about the usage of the term. We who are not parties to the basic regional understanding had been trying to categorize this kind of entity in terms of a single principal attribute rather than in terms of a multiattribute feature for which we had no pigeonhole, no term, no concept.[14]

It is clear in the illustration of the Atlantic coast swamp that the consensus of opinions on its attributes is arrived at in the context of a relatively small group of local people. The understanding is intellectual and emotional, and it is rooted in feeling that accounts for the anger and frustration Burrill observed when the concepts were challenged. This seems to be characteristic of the way communication systems are learned.

An excellent exposition of what is currently known about the communication process is provided by Alton DeLong who advocates use of the communication process as a model for man-environment relations.[15] He

states that within small groups arbitrary codes are developed that simplify and reduce the complexity of the behavioral environment so that effective communication can take place. However, effective communication is limited to the context of the small group in which it is learned. In the examples given previously we saw that the meanings of the symbols (words) may be changed, even within the same linguistic community, or the symbols may change, as in a different linguistic area. But language is only one of the communication systems. "Birdwhistell estimates that of the total social meaning involved in a given transaction no more than 30-35% is communicated verbally."[16] Other nonverbal systems, such as kinesics (body gestures) or proxemics (use of space), pose even greater difficulties for communication between different groups because they are generally out-of-awareness. DeLong declares that "it is quite possible that the limitations we continually face in subcultural and cross cultural communication are the result of limitations built into our central nervous system."[17] Support for this statement is found in the similarities underlying coding behavior, the neurological organization of the brain, the acquisition of intelligence, and the expression of emotion.

> It is through the communication process, internalized relatively early in life, then, that the organism establishes a way of relating to both his physical and social environment. And the specific manner in which he does so irretrievably marks him as a member of a group, a group to whom his allegiances are conservatively drawn, emotionally reinforced, and neurologically guaranteed.[18]

People may not be nearly as flexible as we generally imagine if DeLong is correct. And his analysis does seem to be in accord with many findings noted in the following discussion. This indicates that many different points of view must be considered in environmental planning and design. We must learn to recognize the limitations communication systems impose. "It entails the careful consideration of man's inherent weaknesses as well as his expanding potential."[19] A similar concern with a broad, evolutionary (where are we going), and ecological (how can we survive) perspective is seen in Esser's strategy for research in people-environment systems.[20]

An evolutionary perspective

Aristide Esser, a social psychiatrist, states that there are two main distinctions between human beings and animals. The first is that human beings, in contrast to animals, value the individual. The second distinction is that we have developed culture precisely because we value individual contributions. A problem arises because the cultural environment built up over the last 30,000 years goes very much against natural ways of life and behaviors developed earlier than 30,000 years ago. Esser tries to tie this into the most

recent conceptualizations of the different brains within our central nervous system (CNS). "The built-in difficulty in our CNS is that the process of individuation is guided by the intellectual part of the brain, the cortex, while our early social life is guided by the limbic system, the emotional part of the brain."[21] The total system as conceived by Esser is reproduced in Figure 1.3. Here the limbic system is seen as contributing to the *social brains,* which are derived from shared experiences primarily of an emotional nature. "The collective social brain serves small groups, it fosters knowledge of processes based on allegiances and bonds of blood and soil only. It implies that you value where you are born and to whom you are born, that you know and like your territory and your friends, that the unknown outside is hostile and all outsiders are strangers."[22] In contrast, the *prosthetic brains* are represented as the collective outcome of people's intellectual neocortex. This is based on intellect, which "deals primarily with abstract symbols, not emotional experiences, and as such fosters knowledge not so much of emotional experiences but of patterns, distinctive structures, thereby allowing for interactions based on voluntary agreements instead of loyalties fostered by the communality of blood and soil."[23] The accomplishments of the prosthetic brain are cumulative. More and more explicit conceptualizations of the environment can be built up. But the social brain lacks this advantage, for social learning has to be experienced to be understood. Furthermore, the limbic system is much slower. While an intellectual image can be processed almost instantaneous-

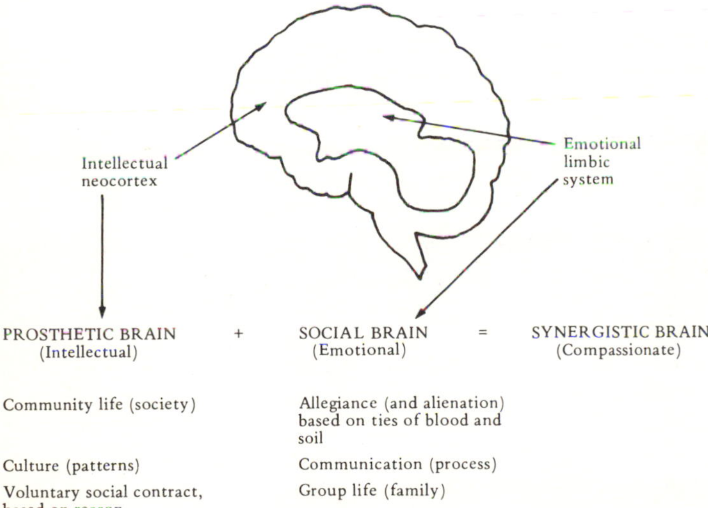

Fig. 1.3 A schema depicting the parts of the human brain responsible for intellect and emotion. (Aristide H. Esser, "Strategies for Research in Man-Environment Systems," in *Behavior, Design and Policy Aspects of Human Habitants,* ed. W. M. Smith, University of Wisconsin, Green Bay Press, 1972. Adapted by permission.)

ly, an emotional image, as, for example, to be friends or to sympathize with someone, takes some time to work through. This leads to conflicts between the ways our intellectual and emotional brains relate to the environment. Esser proposes a new term, "social pollution," to refer to the problems resulting from the clash between the newly created cultural images and the earlier animal images, and suggests that his concept serve as a focus for environmental research. "When we have learned to unscramble the animal and human images in our mind and have learned to project the function of each into the making of particular decisions, social pollution will become preventable. The mastery of simultaneous social and prosthetic brain functioning might be called the synergistic brain."[24] This is represented in a final portion of the system in Figure 1.3. With the operation of *synergistic brain* we will be at one with our environment, and social conflicts between separate groups, tribes, and cultures will cease.

Esser's system is useful for putting the people-environment problem in evolutionary perspective and indicating the goal or direction in which we should evolve. This brings us back to the idea expressed in the opening quotation by Robert Sommer and the introductory paragraphs of this chapter. There is no doubt that we ourselves are creating the social and physical environment in which future generations will evolve. The wisest course for us would be to plan in conjunction with a set of goals that would enhance the development of what Abraham Maslow has termed "the farther reaches of human nature."[25]

According to Maslow, human beings have an evolutionary series of needs that can be arranged in a hierarchy of priorities.[26] When the needs having the greatest potency are satisfied, the next needs in the hierarchy emerge and press for satisfaction. At the most elemental level are the physiological needs, such as hunger and thirst. Once these needs have been met, safety needs emerge, and, in turn, other needs, such as affection, esteem, self-actualization, acquisition of knowledge, and aesthetic fulfillment. In previous periods our basic physiological needs were the main driving force, and in some parts of the world they remain the major preoccupation. But today in places where satisfaction of some of the lower-order needs is possible, a new preoccupation with higher-order needs has developed. To date, we have not learned how to provide adequately for the satisfaction of such needs. The ultimate aim of research in environmental perception and behavior, when viewed in the evolutionary perspective, is to learn enough about people-environment interactions to be able to plan a world in which the physical environment would facilitate satisfaction of the higher- as well as the lower-order needs. It is by satisfying the higher-order needs that we become more fully human, for while the lowest-order needs are shared with all other living things, the higher-order needs are exclusively human.

Maslow's framework, while very suggestive, is difficult to apply directly. For example, we do not know how the higher-order needs may be best

served by environmental manipulations. Some of the studies reviewed in the chapters ahead have grappled with this problem. Their findings provide insights that should help further the pursuit of the solutions to our problems.

Notes

1. Department of State, *Documents for the U.N. Conference on the Human Environment,* Stockholm, June 5-16, 1972, National Technical Information Service, Springfield, Va., 1972.
2. John Kirtland Wright, *Human Nature in Geography,* Harvard University Press, Cambridge, Mass., 1966, p. 10.
3. *Ibid.,* p. 10.
4. Joseph Sonnenfeld, "Geography, Perception, and the Behavioral Environment," paper presented at AAAS Dallas meeting, December 27, 1968.
5. *1965 Directory of Behavior and Environmental Design,* Research and Design Institute, Providence, R.I., 1965.
6. Harold M. Proshansky, William H. Ittelson, and Leanne G. Rivlin, *Environmental Psychology: Man and His Setting,* Holt, Rinehart and Winston, Inc., New York, 1970.
7. Roger G. Barker and Herbert F. Wright, *Midwest and Its Children: The Psychological Ecology of an American Town,* Row, Peterson and Company, Evanston, Ill., 1954, p. 7.
8. Roger M. Downs, "Geographic Space Perception: Past Approaches and Future Prospects," *Progress in Geography,* 2 (1970), Edward Arnold, publishers, Ltd., London, 65-108. Reprinted by permission.
9. *Ibid.,* p. 84.
10. Benjamin Lee Whorf, "Science and Linguistics," *Technology Review,* 44, 229-231, 247, 248, reprinted in Eleanor E. Maccoby, Theodore M. Newcomb, and Eugene L. Hartley, *Readings in Social Psychology,* Holt, Rinehart and Winston, Inc., New York, 1958.
11. Forrest R. Pitts, "The Japanese Landscape as Mood and Intensity," unpublished manuscript, 1960; a shorter version appeared as "A Mirror to Japan," *Landscape* (1960), 9.
12. Meredith F. Burrill, "The Language of Geography," *Annals of the Association of American Geographers,* 58 (1968), 1-11.
13. *Ibid.,* p. 4.
14. *Ibid.,* p. 4.
15. Alton J. DeLong, "The Communication Process: A Generic Model for Man-Environment Relations," *Man-Environment Systems,* 2 (1972), 263-313.
16. *Ibid.,* p. 288.
17. *Ibid.,* p. 280.
18. *Ibid.,* p. 283.
19. *Ibid.,* p. 305.
20. Aristide H. Esser, "Strategies for Research in Man-Environment Systems," in *Behavior Design and Policy Aspects of Human Habitats,* ed. W. M. Smith, The University of Wisconsin—Green Bay, Green Bay, Wisc., 1972.
21. *Ibid.,* p. 2.

22. *Ibid.*, p. 2.
23. *Ibid.*, p. 2.
24. Aristide H. Esser, "Social Pollution," *Social Education*,35, No. 1 (January, 1971), 16.
25. Abraham Maslow, *The Farther Reaches of Human Nature*, Viking Press, New York, 1971.
26. Abraham Maslow, *Motivation and Personality*, Harper & Row, Publishers, New York, 1954.

Two

Personal space and room geography

"Space speaks." E. Hall

The image of the astronaut in a space suit, protected within a space capsule flying through a vast, unknown, and dangerous environment, cannot fail to capture our imagination. It illustrates dramatically how successfully today's technology can be directed to achieve what formerly seemed impossible goals. But the person in the spaceship also serves as an illustration of a finely attuned human-machine system. In microcosm it reflects the kind of concern of most studies in this book, that of creating an environment in harmony with human needs and goals. The astronaut was specially trained and the space capsule carefully designed so that both would fit together to reach a particular goal. The combination represents the culmination of a type of research in human-machine relations known variously as engineering psychology, human factors, human engineering, human factors engineering, and ergonomics.[1] As one of the earliest lines of research at the scale of personal space, it illustrates the sequence of concerns in planning the environment for human beings. Generally, the physical factors are the first to be considered, but later, as the inadequacies of this approach are realized, social and psychological factors are emphasized.

The sequence of studies in this chapter follows a similar order. From concern with the largely physical factors—such as, human limitations in

vision, motor skills, reaction times, and audition—the discussion progresses to an example of industrial design, which also emphasizes some cultural and psychological factors. Later the chapter focuses on the cultural and psychological factors. We will consider first some rather general frameworks of value in sensitizing us to broad dimensions of human use of space. Later we will review more specific studies of how space is structured for particular purposes. These will provide a clearer notion of some of the psychological dimensions of space that should be considered in planning seating or other furniture arrangements within rooms. No easy formulas are provided, as it is clear that situational, temporal, and personality factors are all important and may, singly or in concert, change the psychological definitions of space.

Human factors engineering

Engineering psychology, or human factors engineering (the terms most commonly used for this field in the United States), was born during World War II as deficiencies in design led to the recognition of a need for greater psychological inputs into weapons design.[2] Data from experimental psychology and new research were applied in the design of instruments, radar displays, control systems, aircraft cockpits, and similar components of military equipment. Ironically, the greatest impetus came as the war ended, and the investment by defense industries has continued to accelerate to the present day.

In a very general sense, three historical periods can be distinguished in work on human-machine systems.[3] During the earliest phase the research centered on the machine. People had to adapt to the machine as best they could, and operators were selected to serve the demands of the machine. However, as the size and complexity of machines increased, so did their cost. With airplanes worth millions of dollars, the potential cost of human errors became enormous. When this was realized, the second phase of research began, and it centered on people. The aim was to modify the machines to fit human limits. It was only in the late 1950s that the current emphasis came. Then began the third phase, which centered on the system. The characteristics of both human beings and the machine were and are considered together and adapted to each other, with the total performance of the system as the major goal. Unfortunately, the human being is often considered merely a cog in a larger machine, and the ultimate aim of the system is to advance military rather than long-range human objectives.

Because of the strong military slant noted previously, more is known about human behavior in strange, extreme, and unusual environments than in everyday surroundings. The effects of zero gravity, lunar gravity, and increased gravity have been studied in great detail. Studies of human reactions to the small spaces within space capsules and submarines have

received much research support. Yet we know practically nothing about human response to ordinary spaces of similar size, such as the living room, bedroom, or bathroom. The knobs and dials and seats in spacecraft cockpits are contrived to fit human fingers and form while too many kitchens remain outsized and unsuited to the physical dimensions of the people who use them. The studies cited in this chapter concentrate more on what is known about the behavior patterns in ordinary small spaces. It is not so much a complete summary of the little that is known as a series of examples of the types of approaches that could prove useful in exploring our use of such space. The first study cited provides a link from the largely military-centered engineering psychology by illustrating how many of the basic research principles should be applied in industrial design for civilian life. It also shows an increased concern for social and behavioral factors.

The bathroom: an example of industrial design

Alexander Kira's study, *The Bathroom,* departs from the emphasis on extreme environments and focuses on a familiar environmental situation.[4] With a team of researchers at Cornell University, he analyzed the varieties of behavior that take place in bathrooms. They measured heights, reaches, breadths, ranges of movements, and other physical characteristics of people and tabulated various uses of the bathroom and each fixture in order to design the facilities better. Not only did they study the requirements in terms of human physical dimensions and physiological functions, but they also emphasized the importance of cultural and psychological factors.

The functions of each fixture were examined in relation to human physical dimensions. For example, the main functions of the sink, wash basin, or lavatory are hand-washing, face-washing, and hair-washing. Analysis of films of subjects performing each of the activities allowed the construction of such diagrams as Figure 2.1, which shows the range of body and arm movements in washing the face. The grid shows the amount of space necessary for each body movement. From such charts one can deduce the optimum height of the sink above the floor, the width of the basin for comfortable washing, and other similar physical dimensions necessary for designing the fixture. After all such dimensions had been taken into account for each major activity, one possible design for the sink was worked out, as shown in Figure 2.2. A fountain-type of water source is used to resolve the problem of interference with a spout. It allows for ease of cleaning and has the incidental advantage of usefulness as a drinking fountain. The form of the container follows closely the activity pattern as measured and charted, with contours that set up a swirling action so that the runoff rinses most of the basin. A raised splash rim at the front edge helps to keep the body dry. The lever-type hand control is simpler and

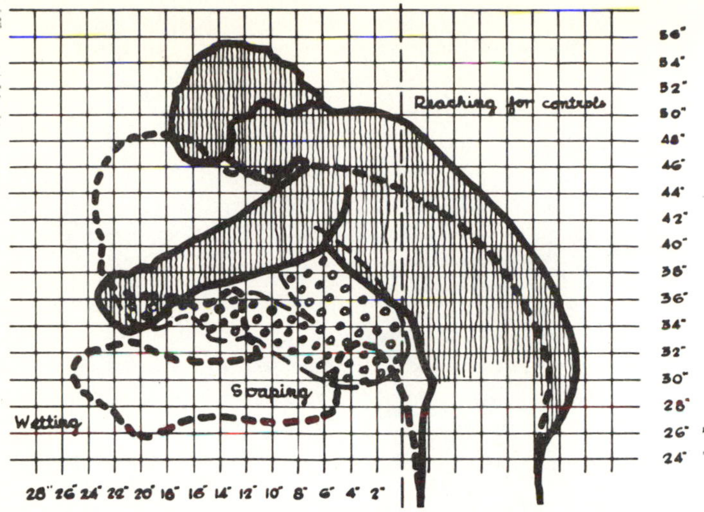

Fig. 2.1 Ranges of body movements while washing the face. (Alexander Kira, "The Bathroom: Criteria for Design," Research Report No. 7, Center for Housing and Environmental Studies, Ithaca, New York, Cornell University Press, Ithaca, New York and Bantam Books, Inc., New York, 1966. Reprinted by permission.)

easier to operate, especially with wet or soapy hands, so it is chosen instead of the common wheel-type lever. In this fashion the design solution aims at achieving the most efficient, convenient, and satisfying fixture for the functions it serves. The same sort of procedure could conceivably be followed to design bathroom facilities for special-user groups, such as the elderly or the handicapped.

However, not all fixtures are as simple as the sink, and often psychological and cultural factors must be considered as well as the physical dimensions and physiological functions. In bathing, for example, the design solutions not only take into account efficiency in the cleansing function and ease in getting in and out but also comfort configurations for the important but often overlooked relaxation function of the bath.

A serious soiling problem in most residential bathrooms is presented by the failure to provide a separate fixture for male urination. The degree to which psychological, cultural, physical, and physiological dimensions are intertwined in a design solution is clearly indicated in Kira's description of this problem. Noting the inability of the male to predict or accurately position the initial point of the urine stream due to temporary and unnoticed dermal adhesions of the urethral opening, Kira recommends a separate fixture, whose specifications are carefully considered.[5]

Ideally, each separate fixture should be carefully designed, and all the fixtures must fit together to provide a total bathroom facility. The final solution should take into account a number of seldom-noticed uses of the

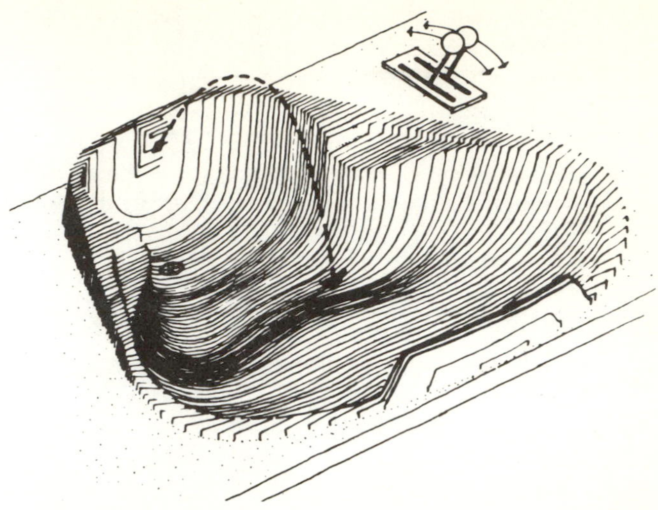

Fig. 2.2 Experimental lavatory incorporating design criteria. (Alexander Kira, "The Bathroom: Criteria for Design," Research Report No. 7, Center for Housing and Environmental Studies, Ithaca, New York, Cornell University Press, Ithaca, New York and Bantam Books, Inc., New York, 1966. Reprinted by permission.)

bathroom, pointed out by Kira's team, such as a place for reading; a refuge for sulking, crying, or daydreaming; a private place for certain hygiene functions; or a private place for other purposes. The attitudes and values of people will of course be a major factor in determining whether emphasis is to be placed on the more luxurious, sensual Roman-bath-with-exercises approach or the more strictly functional system; or whether mental well-being as well as physical well-being and safety is considered.

In America the standard bathroom is a carryover from earlier periods when it was obviously conceived of as a minimum convenience. There has been little change in the past forty years. In part, this results from the organization and attitudes of the plumbing industry and, in part, from American attitudes toward personal hygiene. The plumbing industry does not generally conceive of the total bathroom in design. Instead, plumbers assemble in the field a variety of independently produced and often unrelated items for construction of the bathroom. Attitudes toward personal hygiene also prevent us from giving the problem the attention it deserves. Embarrassment at being considered so bathroom-conscious by the rest of the world may consciously or unconsciously keep us from studying the problem in depth.

Kira's work on the bathroom illustrates clearly how bathroom design can benefit from intensive study of the key physical and physiological factors. There is little doubt that the design of kitchen facilities or other rooms

would benefit from such intensive research on human use and physical dimensions. Kira also recognized that social and psychological factors were important in bathroom design, as, for example, in provision for privacy. One reason such social factors must be considered is that rooms and their furnishings may become infused with social meanings. This will be illustrated immediately by reference to Goffman's sociological interpretation of room use and later as we discuss territoriality within rooms.

The room as a stage

It might be fruitful for designers to consider the problems of room geography in terms of the perspective of Erving Goffman.[6] He examines the way individuals adopt a façade or front that enables them to project to the audience the role they are playing. Part of the façade is one's personal front, which consists of "insignia of office or rank; clothing; sex, age, and racial characteristics; size and looks; posture; speech patterns; facial expressions; bodily gestures; and the like."[7] But, in addition, there is the "setting," which involves the furniture, décor, physical layout, and other background items that supply the scenery and stage props for the performance. The place where the performance is given is referred to as "front region." Here one must behave with suitable politeness and decorum to provide the desired appearance. However, the maintenance of a façade can be a strain, which is relieved by the presence of "back regions" or backstages, which provide places to relax, drop the front, and step out of character. Here, well shielded from those for whom one is performing, one can do all the things that elsewhere might result in negative sanctions.

Perhaps the main shielding places in Anglo-American society are bedrooms and bathrooms, whose important privacy functions have already been mentioned. But front and back regions are found in every sort of human setting and among people of every class. Architects and designers, aware of the types of performances staged in various rooms, buildings, and public spaces are more able to provide appropriate props and back regions, so that architecture absorbs part of the strain of maintaining a façade.

Proxemics

Aspects of a setting—such as furniture, décor, and physical arrangements—are invested with social meaning, and, in addition, space itself is assigned meanings. The way we use space may be regarded as a communication system learned in the context of a small group. Members of the same group use interpersonal space in a similar fashion, while other groups may assign other meanings and spatial distances. Each culture structures the spatial behavior of its members, but we are rarely aware of it. Though

much of the communication within a culture is on a nonverbal level, the messages are clear to the members, for they understand their own group's language of behavior, or "the silent language," as the anthropologist Edward Hall expressed it in the title of his book.[8] This silent language of behavior shows up in innumerable ways in everyday life. In a lively, anecdotal manner Hall's book illustrates how culture operates on a personal level. Most relevant to our discussion here is his contention that "space speaks,"[9] which is treated in greater depth in his later book, *The Hidden Dimension.*

The central theme of *The Hidden Dimension* is social and personal space and our perception of it. Hall coined the term *proxemics,* which he defines as "the interrelated observations and theories of human use of space as a specialized elaboration of culture."[10] He contends that people from different cultures speak different languages and inhabit different sensory worlds. This is reflected in the variable use of space by people of different cultures.

Proxemics has three aspects: fixed-feature, semifixed-feature, and informal. Fixed-feature space is the type of space usually considered in architectural design. It includes buildings, groupings of buildings, and boundaries between rooms. Perception of this type of space is considered in greater detail in Chapter 3 on architectural space. It should be noted, however, that not all aspects of fixed-feature space are necessarily visible. For example, the boundary separating one yard from another in suburbia is real to the owners though not marked by any visible objects. In similar fashion, people carry around with them internalizations of fixed-feature space learned early in life.

Semifixed features include chairs, tables, and other furniture that can be moved around within a fixed-feature space. Which features are fixed and which are movable may vary with the culture. For example, in Japan some walls are movable to provide changing spaces for the changing activities of the day. In some places chairs may be fixed rigidly while elsewhere they can be adjusted at will to provide whatever seating configuration may seem suitable.

Informal space includes the distances maintained in encounters with others. The category is called "informal" because it is unstated and for the most part outside awareness. However, informal spatial patterns have distinct boundaries, within which specific types of behavior occur.

Hall and his associates have worked out a detailed set of distance zones, based on observations and interviews with "noncontact, middle-class, healthy adults, mainly natives of the northeastern seaboard of the United States."[11] These may be studied in Table 2.1, which also notes the degree to which each of the senses is actively involved at each distance. While the measured distances are accurate only for the group studied, a similar chart could be constructed for any culture group. Each of the four distances has a near and a far phase. That people accurately perceive and react to the

Table 2.1 Chart Showing Interplay of the Distant and Immediate Receptors in Proxemic Perception

	0′	1′	2′	3′
Informal Distance Classification	INTIMATE Close	Not close	PERSONAL Close	Not close
Kinesthesia	Head, pelvis, thighs, trunk can be brought into contact, or members can accidentally touch. Hands can reach and manipulate any part of trunk easily. Hands can reach and hold extremities easily but with much less faculty than Seated can reach around and touch other side of trunk; not so close as to One person has elbow room. 2 people barely have elbow room; one can reach out and grasp an extremity. Just outside touching distance. Out of inter-			
Thermal Receptors				
Conduction (Contact)				
Radiation	Normally out of awareness			
Olfaction	CULTURAL ATTITUDE			
Washed skin and hair	OK			
Shaving lotion-perfume	OK Taboo			
Sexual odors	variable Taboo			
Breath	Antiseptic OK; otherwise taboo			
Body odor	Taboo			
Smelly feet				
Vision				
Detail vision (Vis. ∠ of fovea 1°)	Vision blurred distorted	Enlarged details of iris, eyeball, pores of face, finest hairs		Detail of face seen at normal size; eyes, nose, skin, teeth condition, eye lashes, hair on back of neck
Clear vision (Vis. ∠ at macula 12° hor., 3° vert.)		25″ X 3″ on eye, nostrils or mouth	3.75″ X .94″ upper or lower face	6.25″ X 1.60″ upper or lower face
60° scanning		1/3 of face; eye, ear, or mouth area, face distorted	Nose projects whole face; seen face undistorted	Upper body, can't count fingers
Peripheral vision		Head against background	Head and shoulders	Whole body movement in hands—fingers visible
Head size		Fills visual field far over life size	Over normal	Normal size
Note: Perceived head size varies even with same subject and distance				
Additional notes	Sensation of being crosseyed.			
Tasks in submarines		67% of tasks in this range		23% fall in this range
Artists observations of grosser			Very personal distance	Artist or model has to dominate
Oral Aural	Grunts, groans	Whisper	Soft voice. Intimate style	Conventional
Note: The boundaries associated with the transition from one voice level to the next have not been				

SOURCE: Edward T. Hall, *The Hidden Dimension*, Doubleday & Company, Inc., Garden City, N.Y., 1966, pp. 126-127. Reprinted by permission of Lurton Blassingame, the author's agent.

4′	5′	6′	7′	8′	9′

SOCIAL-CONSULTIVE

Close	Not close
above. result in accidental touching. ference distance; by reaching, one can just touch the other.	
Animal heat and moisture dissipate (Thoreau)	
— Taboo	
Smallest blood vessels in eye lost. See wear on clothing, head hair seen clearly	Fine lines of face fade. Deep lines stand out. Slight eye-wink, lip-movement seen clearly
10″ X 2.5″ upper or lower face or shoulders	20″ X 5′ 1 or 2 faces
Upper body and gestures	Whole seated body visable; people often keep feet within other person's 60° angle of view
Whole body	Other people seen if present
	Normal to beginning to shrink
A picture painted at 4′–8′ of a person who is not paid to "sit" is a portrait.	Too far for a conversation
modified voice. Casual or consultive style precisely determined.	

distances is indicated by the behavior that regularly recurs within each zone and that provides their descriptive names, i.e., intimate, personal, social-consultative, and public distances.

Intimate distance is readily perceived by the intense involvement of all the senses. The other person is heard, seen, smelled, and felt. Physical contact or the high possibility of physical involvement is uppermost in the awareness of both persons. This is the distance of lovemaking, comforting, and protecting. It is not considered appropriate in public by Americans, although this rule is not shared by Russians or Middle Eastern peoples, according to Hall's observations. He also notes that defensive tactics are used in such crowded places as subways and buses to remove the real intimacy that the close proximity might suggest. Thus the eyes are not fixed on the other, the hands are kept at the side, and physical contact is avoided if possible. This is in keeping with the hypothesis of Argyle and Dean, who suggest that intimacy is a function of several interrelated variables, including physical proximity, eye contact, the nature of the conversation, and the amount of smiling.[12] As one of these components is altered, one or more of the other components are shifted in the opposite direction to maintain equilibrium. In the case of crowded places, where proximity is forced, the lack of eye contact, smiling, and conversation helps shift the balance away from intimacy. This illustration shows the importance of considering a particular behavior in terms of context. One must understand the entire system to interpret correctly the meaning of a particular aspect of behavior.

Personal space is the minimum spatial separation most consistently maintained. It could be considered as a small portable territory or invisible protective bubble within which encroachments are generally unwelcome and often distressing. Personal space is not necessarily spherical in shape, for people can tolerate close presence of a stranger at their sides more readily than directly in front. A study by Horowitz, Duff, and Stratton showed that people would approach closer to others from the rear or side than the front and closer to a woman than to a man.[13] At personal distance subjects of personal interest and involvement can be discussed. The voice level is generally moderate, and the body smells and heat are not generally noticeable. The far phase of personal distance is the point where two people can touch fingers if they extend their arms. Thus it represents the limits of physical dominance, for beyond it a person cannot easily touch someone else.

Social distance is more commonly used for impersonal business. The close phase tends to be used at casual social gatherings and is generally maintained by people working together. Speech is at the normal conversational level. For business of a more formal character, the far phase is more common. The distance across the large desks of important people places visitors at the far phase of social distance. Hall observes that this distance is suitable for screening people from each other. For example, receptionists in outer offices can continue working without seeming rude even though

others may be present as long as the far phase of social distance is maintained.

Public distance is associated with several shifts in behavior. The voice must be louder, and changes in the choice of words or phrasing usually occur. It is the distance of public speakers or actors. Subtle shades of meaning conveyed by the normal voice are lost as well as details of facial expression. Gestures and body stance must be exaggerated to communicate clearly.

Hall's observations on the four principal distances and categories of relationships associated with each indicate how culture can condition our perception and use of space. Clearly, human space requirements should not be thought of as simply the amount of air displaced by our bodies. Planners and architects should bear in mind these zones and their meanings in designing rooms, buildings, and cities to serve their purposes yet avoid unnecessary stress. When planning is done on a world scale, the differences in spatial zones become extremely important. Buildings spatially appropriate for one culture may create unanticipated proxemic problems in another culture.

Hall's bold generalizations may sensitize us to the importance of proxemics as a communication system and, by implication, to other systems of communication that operate out of awareness. There has been some confirmation of Hall's observations under controlled conditions,[14] but sample sizes were small, and much remains to be learned about the degree to which various distance zones are valid and how much they vary among cultural and subcultural groups. There are indications that personal space and spatial behavior may vary with characteristics of the physical environment—such as, room size, number of occupants, table and chair availability, shape of table, and location in reference to the speaker—as well as with characteristics of the individual—such as, enduring personality traits, age, sex, and momentary feelings—with characteristics of the task of relationship—such as, cooperation or competition, conversation, friendship, or shared attitudes—and with characteristics of the other individual—such as, leadership, attraction, and stigma.[15]

Spatial invasion

Just how important the individual cushions of space are and how they vary with the situation and the individual can be gauged by glancing through the discussion of spatial invasion by Robert Sommer.[16] He notes, for example, police awareness of the distressing effects of violation of personal space as exemplified by their standard interrogation technique. By placing their chairs too close to the subjects law enforcement officers can maintain stress and undermine the confidence of the persons. The amount of space desired will vary among people and situations. People who drive au-

tomobiles are no doubt familiar with the uncomfortable sensations that arise from being tailgated on the highway. Introverts keep people at a greater distance than do extroverts; and schizophrenics, lacking stable self-images and clear self-boundaries, may come too close or remain too far for normal conversation.

Perhaps the most direct way to explore individual distance and social space is to approach people and observe their reactions. This was the method employed by Sommer, who uses an effective combination of techniques, namely, naturalistic observation and experimentation supplemented by short interviews. In a series of experiments in real-life settings—such as, libraries, bus stations, airports, cafeterias, and other public places—he and his associates investigated how people try to preserve their privacy by selecting, marking, and defending small territories. The first step in such research is direct observation in the natural setting. An unobtrusive observer notes the spatial behavior of individuals in relation to the physical setting and to each other. On the basis of these observations, some tentative hypotheses can be tested by direct experimentation, as in the library situation where the researcher entered the library reading room and sat too close to others. Additional information was obtained by asking people to indicate on a diagram of a library table and chairs where they would sit to keep away from others or to actively discourage others from approaching. Such research has the great advantage of leaving the real-life situation totally undisturbed until observations have been made of how behavior operates in the particular setting. Only at this point are small changes introduced by the researcher who becomes an actor in the situation. His presence is natural enough so that the subjects intruded upon react as they usually would in such circumstances. The use of diagrams to mark seating positions enables the researcher to quickly examine changes that might occur with further variations. It also provides extra evidence to corroborate or contradict the data obtained by direct observation.

The studies of spatial invasion in libraries carried out by Sommer and his associates indicate a variety of strategies for safeguarding personal space. However, it was observed that many of the difficulties of defending this invisible space may be overcome by position, posture, and gesture. Position refers to a person's location in a room. Choice of an area that conveys a clear meaning is important. A central location differs in this respect from a corner, alcove, or side area hidden from view. Posture refers to a person's stance. One can aggressively spread out one's belongings as markers to control a space or pull oneself in to take up little space. Gesture can be used to threaten or discourage a would-be invader or to signal, by facial expression or eye contact, that one is ready for company. Sommer found that whether students were asked to diagram a retreat or diagram a defense strategy for maintaining a space made a great difference as to the positions selected within the room and at a particular table. Students wishing to sit as

far as possible from other people would overwhelmingly select end chairs, while those wanting to keep others away from their table would almost unanimously choose the middle chair.

Curiously enough, flight was the most common reaction to an invasion of personal space in studies of a mental hospital and a college library. The person might make defensive gestures, shift posture, or move away, but if the intruder remained or continued to advance, the victim eventually fled. Almost never was there a direct verbal response to the invader. If a more permanent territory were invaded rather than personal space, this probably would not be the pattern, as is indicated in the discussion of territoriality in a mental ward found in the next chapter and in the reactions of residents of a Welsh old people's home noted by Lipman.

Chairs as territory

Alan Lipman studied chairs as territory in old people's homes in South Wales.[17] In private homes, clubs, pubs, or university classes, it is common for specific chairs to be used regularly by the same people, even though this is not official or even mentioned. Most of us have have probably been miffed at one time or another on arriving somewhere to find that our regular seat has been usurped. In an old people's home, where residents generally spend most of their waking hours seated, this type of pattern is developed to the extreme. Lipman logged the proportion of time each chair was occupied by the "owner" as opposed to another person. The percentages may be seen on the plan of the sitting room (Figure 2.3). It is clear that there was a rigid maintenance of personal chair preferences even in the face of considerable physical discomfort and an official policy, frequently stated, that no chair belongs to any particular resident. On sunny days certain chairs become almost unbearable because of the heat, and, on overcast days and in evening light, seats were too poorly illuminated to allow reading without strain. Yet the stoicism or passive acceptance was such that even at these times no attempts were made to move to vacant chairs to more favorable locations in the room. The television set in the upper right corner could not be viewed readily from Chairs 15 to 18 and 23 to 27. But prolonged maintenance of uncomfortable physical positions by people in these seats seemed preferable to changing to empty chairs for more comfortable viewing. The profound psychic and emotional significance of the chair ownership to the occupants was demonstrated by an occasion when someone sat in another person's chair. The owner would verbally attack the other, and the usurper would relinquish the seat. The passion that could be aroused is indicated in the following incident described by Lipman:

> Once Mrs. M., a small and frail woman of 90, found the contentious chair No. 25 occupied by Mrs. A. (the wife who saw it as her reserve chair in this room). Mrs. M.

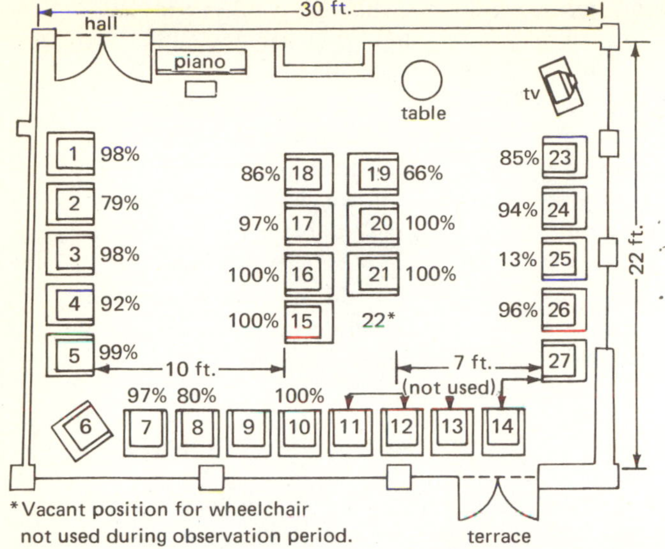

Fig. 2.3 Percentages of time spent in particular seats. (Alan Lipman, "Chairs as Territory," *New Society*, 1967, IX, 565. Adapted by permission.)

demanded, on her return from the toilet, that the chair be vacated. Mrs. A. refused, claiming the chair was hers. After some argument, Mrs. M. lifted her walking stick and began beating Mrs. A. about the head and shoulders. In the uproar of weeping, cries and moans which followed, the opinions expressed by the other residents fell into two categories. Some saw Mrs. M's behavior as "evil" and "vicious"; the other attitude is epitomised in the comment: "She's quite right. That one's always coming in here when she's got her own place in the chair by the other room. Serve her right, it do."[18]

The situation described by Lipman is admittedly extreme, but it does illustrate a couple of important points. First, it is a clear example of how important to the individual a very small space or territory can become. Second, it shows how profoundly the behavior and experience of such inactive groups can be affected by the physical arrangements of the furniture.

Territoriality in small rooms

Territoriality, as it develops in isolated or confined groups—such as, in space capsules, submarines, or Arctic or Antarctic stations—has been studied extensively by Altman, Haythorn, and their associates.[19] Their studies show that interpersonal relationships are affected by an environmental milieu, such as isolation. But this is not all: In addition, the physical

environment is actively used to manage social relationships in accord with interpersonal compatibility. An example of their approach is an experiment described by Altman and Haythorn that compared the spatial habits of pairs of men socially isolated in a small room for ten days with those of matched, nonisolated groups. The isolated men were matched as closely as possible with the nonisolated men in terms of personality, age, education, and other demographic variables. But while the isolated men lived in a room 12 feet by 12 feet in dimensions, the nonisolated control group lived in barracks, ate at the base mess, and were free to use the recreation facilities of the base. Altman and Haythorn had a controlled experimental situation with the main differences being environmental ones. The behavioral differences observed could then be attributed to the environment variables.

"Territoriality was defined in terms of the degree of consistent and mutually exclusive use of particular chairs, beds, or sides of the table by dyad members."[20] They found that those in the socially isolated condition showed a gradual increase in territorial behavior, as defined, and developed a general pattern of social withdrawal. "Taken together, the territoriality and social activity data suggest that isolates may have begun drawing a psychological and spatial 'cocoon' around themselves, gradually doing more things alone and in their own part of the room."[21] Territorial behavior appeared to develop in a regular sequence. Among the isolates there was an early preference for particular beds, then side-of-the-table territoriality, and finally a marked preference for chairs, which reached its peak during the final days of isolation. It is interesting to note that the bed and side-of-the-table territoriality, where there were fixed positions, developed more rapidly than chair territoriality, which involved movable objects.

The same study also examined the effects of group personality composition on territoriality and social activities. The pairs of men in isolation and nonisolation were matched in terms of personality composition of groups. In general, incompatibility in the traits directly associated with interpersonal matters resulted in high territoriality. The higher territorial levels are not so evident in the more compatible pairs. It appears that territorial structuring was one of the factors used by these people to counteract environmental stress.

Comparison of several similar studies and of the characteristics of pairs who completed the isolation experiment with those who failed was revealing. A pattern was found reflecting similar modes of adaptation by compatible and successful groups versus incompatible and aborter groups. The incompatible pairs and the aborters were very low at first and high later in development of territorial patterns. In contrast, compatible pairs and those who completed the period of isolation started with high amounts of territorial behavior that eventually declined. Altman interprets this as indicating the adaptive nature of territorial behavior: "such behavior early in a

relationship could be taken as a sign that the group members had begun behaving to create a viable relationship with one another."[22]

The eventual decline of territorial behavior among the compatible and complete pairs may indicate that, once it has served its purpose in group formation, strict territoriality may be no longer so essential. The two patterns involve different approaches in the active use of the environment in managing interpersonal relationships. One pattern is anticipatory, involving the prearranging-prestructuring of the environment to create certain interaction settings, while the other is reactive, i.e., a use of the environment in reaction to developing events. Some further studies indicate that the type of architectural plan preferred for a two-compartment undersea capsule depended on the type of relationship anticipated with one's companion. It seems likely that territorial behavior similar to that described here would take place in other settings, such as shared rooms in college dormitories.[23] Probably in most design situations there would not be the same possibility of selection for personality differences as seen in the studies just cited. However, it is important to bear in mind the possible consequences of individual differences in the use of space. This will be noted again at many different scales in the following chapters, as will many other instances of variations in human territoriality. When such behavioral dimensions are better understood, thoughtful design of physical arrangements of rooms may be possible.

Social effects of furniture arrangement

The same rigid type of arrangement of seats side by side and back to back in rows as can be seen in Figure 2.3 was found by Sommer in western Canada in an earlier study, which originally kindled his interest in environmental engineering. Like Lipman, he was studying elderly people in an institution, specifically, an elderly women's ward at a state hospital. Sommer was called in to help discover what was wrong with the place. Although recent expensive renovations had resulted in good publicity, the head physician was dissatisfied with the outcome, for, in spite of improvements in ward appearance, there were no changes in the mental states of the occupants. Not only was there the side-by-side and back-to-back arrangement noted previously, but around several columns there were four chairs, each one facing a different direction. Sommer described the situation:

> The ladies sat side by side against the newly painted walls in their new chrome chairs and exercised their options of gazing down at the newly tiled floor or looking up at the new fluorescent lights. They were like strangers in a train station waiting for a train that never came. This shoulder-to-shoulder arrangement was unsuitable for sustained conversation even for me. To talk to neighbors, I had to turn in my

chair and pivot my head 90 degrees. For an older lady, particularly one with difficulties in hearing and comprehension, finding a suitable orientation for conversation was extremely taxing. I hardly need add that there was no conversation whatever between occupants of the center chairs that faced different directions.[24]

In Wales and western Canada, and probably in most other institutions as well, the furniture was considered as something to be arranged by the staff for their convenience rather than as a potential therapeutic tool. It is easier to sweep, clean, or move along straight rows than around groups or clusters of chairs. That the furniture was arranged for the convenience of the staff became apparent from their complaints and sarcastic comments in resistance to experimental changes as they were proposed and initiated. The contrast between healthy and institutionalized people was also apparent: during visiting hours the families and friends of the patients rearranged the chairs into small groups so that they could face one another and converse easily. In other words, they arranged their environment to suit their needs, while the more passive patients were arranged by their environment. Furthermore, no one had thought of asking the elderly women how they would like the furniture arranged; thus the changes were planned and initiated by people who spent no time on the ward.

In search for a solution to this problem, Sommer started by observing seating arrangements in such places as homes, bus depots, railway stations, cafeterias, theaters, and hotel lobbies. He noticed that very little interaction occurred where seats were placed in long lines and that the most usual arrangement for conversations in cafeterias seemed to be across the corners or in face-to-face positions across tables. Eventually it was decided that the women would be more likely to converse if they sat facing one another rather than shoulder to shoulder, and that interaction might be facilitated if the large, open areas were broken down into smaller spaces. This was accomplished by simply adding a few tables and rearranging the chairs around them.

Initially there was resistance, not only from the staff, whose route through the ward was broken up, but also from the women themselves, who continued to move chairs back to their former places along the walls, even for some years afterward. But making the tables attractive with flowers and magazines gradually induced the women to use them. Comparison of conversations before and just two weeks after the changes indicated that already both transitory and sustained interactions had increased markedly. In addition, there was a remarkable increase in the amount of reading, and, as a further unanticipated outcome, an increase in craft activities throughout the day. Thus changing a single element induced changes throughout the social system. The same sort of effect is seen again in Chapter 3 when changes were made in the psychiatric ward of a large metropolitan hospital.

Seating position and participation

Systematic study of spatial factors in small groups is a recent development. In his book *Personal Space,* Sommer provides a comprehensive review of the present state of knowledge in people-environment relations at this scale and traces the start of such research to a study by Steinzor in 1950.[25] According to Steinzor's account, this began almost accidentally while he was investigating another aspect of behavior in face-to-face groups. In one discussion group a participant was observed changing his seat so that he could sit opposite another member with whom he had had an argument. This action led Steinzor to develop the hypothesis that "seating arrangement in a small face-to-face group helps to determine the individuals with whom one is likely to interact."[26] Reexamining the records of several sessions of group interaction, he found that this was true. People sitting opposite each other at a round table followed each other in comments significantly more frequently than would have been predicted. Thus it seemed that individuals were more likely to be stimulated by the remarks of those persons they could more completely observe. An example of the kind of results such effects might produce is provided by a study by Strodtbeck and Hook who recorded seating arrangements in experimental jury sessions in Chicago.[27] The jurors met in a jury room containing a rectangular table with five chairs on each side and one at the head and foot. The jurors' first task was to select a foreman. The tendency was to select one of the persons seated at the head or foot of the table, the people most clearly visible to the majority of jurors. Interestingly, it was found that the initial choice of seats was not random, for the higher-status people were found in head chairs more often than would be expected by chance. This was confirmed by Sommer's study.[28] Working with discussion groups in a cafeteria setting, he demonstrated that leaders tended to select the head position at rectangular tables. Other members of the group would then arrange themselves so they could see the leader. Here is seen a clear link between table territoriality and dominance.[29] Other social psychologists have indicated the importance of the physical arrangements and communication channels.[30] Group leadership is closely correlated with the person's position in the communication net, as is the degree of satisfaction. Subjects in the most central position enjoyed their work most; those in peripheral positions enjoyed it least.

The importance of individual choice must be considered in the relationship between participation and location. It has already been noted that the leaders tend to select dominant positions at tables. Similarly, in a classroom students may select seats according to the degree to which they wish to participate. Sommer and some of his associates mapped the degree of participation in straight-row classrooms by dividing each row into separate sections. Figure 2.4 shows that there are systematic differences in the degree of participation depending on location. Those in the front row and

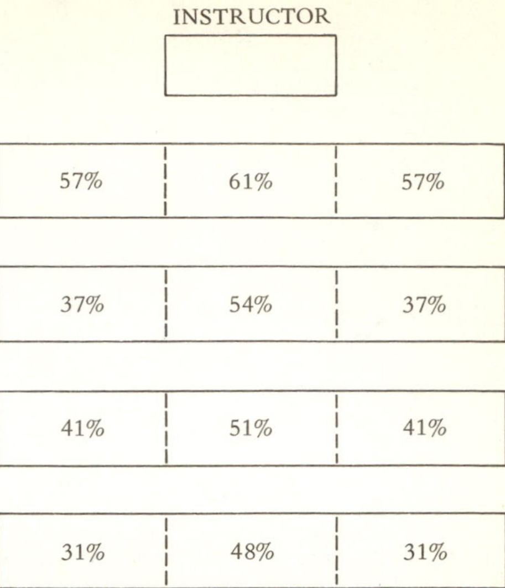

Fig. 2.4 Ecology of participation in straight-row classrooms. (Robert Sommer, *Personal Space*, Prentice-Hall, Inc., Englewood Cliffs, New Jersey, 1969, p. 118. Reprinted by permission.)

in the centers of rows closer to direct eye contact with the teacher participated more frequently. However, in cases where the choice of seats was not voluntary, so that interested students could not sit where there was maximum visual contact with the teacher, the connection between locations and participation was less clear. In a large lecture hall divided into a large center section and two smaller side sections, almost half the questions came from those in the first two rows of the center section. Another third came from students sitting along the side aisles, but there was very little participation from the "faceless mass" in the center.[31] This is reminiscent of Bruno Bettelheim's observations in the grisly context of World War II concentration camps. One means of surviving in the camp was to remain inconspicuous and therefore unnoticed. The desire for anonymity was so great that "during morning roll call a fight of one against all often began for the least visible positions in the parade ground formation." Those in the front, back, and side rows were the most likely recipients of kicks or blows. The preferred positions were in the center, where protection was provided by prisoners on all sides. These examples, as well as the Altman and Haythorn studies of territoriality, indicate the interactive nature of the human-environment system at the scale of a room. Location is important, and this is recognized by the participants in a setting who place themselves in the position that best serves their purposes. The functional significance of position in relation to furniture and to other people is further illustrated

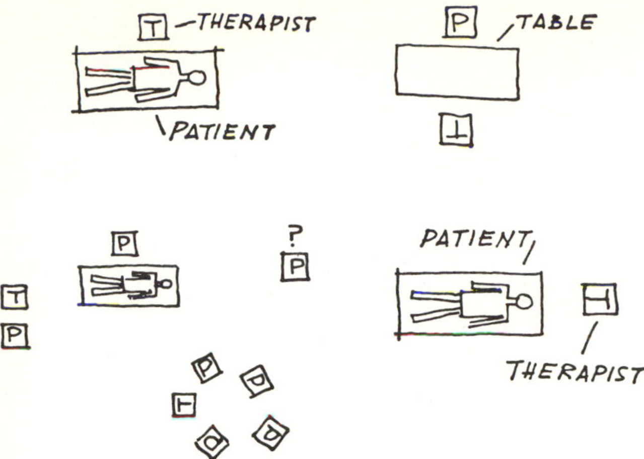

Fig. 2.5 Seating arrangements for psychotherapy. (Paul Goodman, *Utopian Essays and Practical Proposals,* Random House, Inc., Alfred A. Knopf, Inc., New York, 1964, pp. 158-161. Reprinted by permission.)

by an example that explores the range of meanings possible in the arrangement of the patient and therapist for psychotherapy.

Seating arrangements within rooms and their significance in functional planning is the topic of a stimulating discussion by Paul Goodman.[32] His essay indicates the rich variety of psychological meanings implicit in seating arrangements for various activities: psychotherapy, religious services, eating, democratic legislatures, universities, and theaters. In psychotherapy, for example, with only two seats involved, the patient's and the therapist's, one might assume there is little room for variation. Yet four radically different plans are utilized by four important schools of therapy. In each the seating arrangements are altered according to the theory and aims of the school (Figure 2.5).

In Freudian therapy the aim of treatment is to bring to consciousness certain complexes of repressed ideas. To facilitate "free association" of these thoughts and images, the patient lies on a couch with the therapist behind him and out of sight. The seating arrangement of the Sullivan school is face to face across a desk. This treatment is designed for schizophrenics rather than psychoneurotics and appeals to that part of the patient's personality that can respond directly. The content discussed tends to be current events of the day, at the office, with family and friends. These are "laid on the table" and discussed objectively, and the patient is given reassurance and strengthened self-esteem. The method of character analysis of Wilhelm Reich has the patient lying exposed on a couch, naked or nearly so, with the therapist alongside to observe, give directions, or touch if need be. Treatment is importantly physical, for the character

defenses to be overcome are maintained by rigid muscles or other somatic reflexes. To the gestalt psychologists, any fixed seating plan overstructures the present. Their theory of neurosis is that the patient repeats archaic habits to avoid anxiety. The aim is to heighten awareness of the changing present and recover possibilities of creative adjustment to it. Thus the seating is freely altered, and on various occasions the therapist might be unseen or leave the room, the patient and therapist might change places, or there might be a group.

An article by Winick and Holt indicates just how variable the seating arrangements can be in sessions of group analysis.[33] They observed that differences between hard and soft or fixed and folded chairs are psychologically significant, and that brightly colored chairs may at times serve therapeutic purposes. One seat with red upholstery became known as "the hot seat," while a blue one was called "cool." In sessions where patients experienced changes in their moods, spontaneous adjustment of the seating arrangements occurred, such as moving the chairs into a circle in the presence of anxiety, division into subgroups, or withdrawal of individual members to the side, back, or even the floor. Distance, arrangement, and the symbolism of space and furniture are seen as important.

Summary

The studies in this chapter provide an introduction to a number of themes that will be repeated in later chapters. The concern with creating an environment in harmony with human needs and goals was indicated in microcosm at the scale of personal space and room geography. The examples of the astronaut in the space capsule and the design of the bathroom illustrate clearly how the environment can be designed to fit our physical dimensions. The more important question is how the environment can be designed to fit our social and psychological dimensions. This is a more complex task, and one we know little about. It is the task to which most of the research reported in this book is devoted. It requires a deeper understanding of human nature, the nature of the environment, and the nature of people-environment interaction.

The nature of people-environment interaction at the scale of personal space may best be studied by an ecological approach. It appears to operate as a system, as was illustrated by the unanticipated behavioral consequences of the change in furniture arrangements reported by Sommer. Not only did the adjustment of chairs and tables lead to a change in the amount of conversation among the elderly women, but there was also a remarkable increase in the amount of reading and craft activities. Similarly, although physical proximity may generally tend toward intimacy, Argyle and Dean suggest that the balance may be shifted by other components of the situation, such as the amount of eye contact or smiling or the nature of the

conversation. We are just beginning to realize the extent of such baffling interconnections.

The environment, here composed mainly of other people and furniture arrangements within a room, affects behavior but is clearly not a determining factor. The influence of the environment appears to be greater with less flexible people, such as the elderly or infirm. This was noted in the contrast between patients and visitors in an old people's ward in a hospital. While the visitors would rearrange the chairs in order to face each other, the patients by themselves would not. They were arranged by the furniture, instead of arranging the furniture. What happens in extreme form to the elderly, infirm, or inflexible, probably happens to all of us to some degree. This may be especially true in public places, where we may feel we have no right to interfere with the current arrangements. But even where it is possible or permissible to alter the room environment, we may not do so due to a lack of awareness of other alternatives. This raises the important question of education for environmental awareness, a topic discussed in *Design Awareness* by Robert Sommer, and one of great concern in succeeding chapters of this book.[35]

There are probably great individual differences in the ability to manage creatively the environment to serve our needs. Simple physical determinism is not the main process in people-environment relations, for individual choice plays a large role. Some general awareness of the way spatial distances may alter our relationships with other people or the way the physical arrangements may modify our reactions to a room is indicated by individual selection of positions for various purposes. In the studies reviewed, we have seen, for example, how the more successful pairs in the isolation experiment in a small room used territoriality to establish viable relationships with each other. On the other hand, the less successful pairs did not establish territorial behavior until too late. Similarly, some people selected seating positions in accordance with their general desire for dominance or to defend a space, or to facilitate participation in classrooms. In the case of concentration camps, the parade group position was selected to increase survival chances. The room arrangement could also be used as part of a person's façade, and psychiatrists of different schools alter the furniture to fit their theories. All these examples indicate a very active role in our assigning meanings and structuring the environment to meet our needs and purposes. Good design at the scale of the room can simplify our task and release energies for other purposes.

Personal choice plays an important role in people-environment interactions; but we may not be quite so free as we think. The fundamental importance of culture in conditioning our approach to environment was emphasized in the discussion of proxemics. Although this communication system is largely outside awareness, it serves to create a common set of spatial meanings for members of the same culture. But misunderstandings

may arise in situations of cross-cultural contact due to differences in expectations.

An awareness of proxemic patterns may provide some general guidelines for designers, but these must be considered in light of the specific circumstances of the design in question. Some flexibility should therefore be built into the design unless the unique qualities of the setting and of those using it are thoroughly understood. Deficiencies in knowledge may be overcome to some degree by research in environmental perception and behavior. But, given the present state of knowledge, one must still regard each new design as an experiment whose results require careful observation to provide information for the next round of design. This is not an optimistic statement, but at least there is an increased understanding of the degree of ignorance and some notion of where to seek new information.

The question has been raised of whose perceptions are most important as a design consideration. It is evident that users' opinions are often ignored. Thus the physical arrangements of bathrooms may be based on ease of assembly by the plumbing industry, and furniture may be arranged for the convenience of the cleaning staff rather than the patients of an institution. Similar disregard of the prime users may be seen in planning at many different scales in the chapters that follow.

Research reviewed at the scale of personal space and room geography has provided examples of many different types of methodology. These range from casual observation to highly controlled experiments. In some cases the physical measurements of the users were important, but most often it was the behavioral dimensions that were sought. Students could experiment with these techniques in examining their own environment at the scale of the room. Naturalistic observation can be applied in almost any real-life setting in an unobtrusive manner. Some of the preliminary observations in classrooms, bars, cafeterias, and playgrounds could be tested in similar settings or in new situations. The important thing is to arrive with some idea of the type of behavior to be observed and the environmental dimensions along which it is presumed to vary. Interviews have also been used, and these can speed up the process by posing questions about situations that have not yet been observed or that hinge on social meanings or perspectives developed over time, which are not readily observable. Such symbolic aspects of the situation appear at the level of personal space and become increasingly important in broader segments of the environment. The value of the questions and questionnaire may depend on how thoroughly the situation is understood beforehand, and on how carefully they are designed to fit the subjects interviewed. Naturalistic experiments may be easily tried, and they have the advantage of producing real-life reactions. Laboratory experiments require extremely careful design, for they tend to isolate behavior from the natural context, and if this is not

done with sensitivity the results may not be transferrable to the real world. But if successful they also provide great explanatory power of specific variables.

Although the majority of the researchers at this level appear to be behavioral scientists, it is interesting to note that much of the inspiration for study design is derived from field observation techniques of biologists and, in particular, the testing in human situations of concepts based on animal territoriality. This same source of inspiration is seen at many other scales. A lively research atmosphere comes from observation and analysis of behavior as it occurs in real-life situations.

Notes

1. Examples of work in this field may be found in such publications as *Ergonomics, Human Factors,* and *Journal of Engineering Psychology.*
2. A review of the development of research in engineering psychology is provided by Walter F. Grether, "Engineering Psychology in the United States, "*American Psychologist,* 23, No. 10 (October, 1968), 743-751.
3. According to the review of Maurice De Montmollin, *Les Systèmes Hommes-Machines,* Presses Universitaires de France, Paris, 1967, p. 5ff.
4. Alexander Kira, "The Bathroom: Criteria for Design," Research Report No. 7, Center for Housing and Environmental Studies, Cornell University, Ithaca, N.Y., 1966. Also published in a paperback edition, by Bantam Books, New York, 1966.
5. *Ibid.,* p. 142.
6. Erving Goffman, *The Presentation of Self in Everyday Life,* Anchor Books, Doubleday & Company, Inc., Garden City, N.Y., 1959.
7. *Ibid.,* p. 24.
8. Edward T. Hall, *The Silent Language,* Doubleday and Company, Inc., Garden City, N.Y., 1959. Also published as a paperback (Premier Books, New York, 1961), p. 10.
9. *Ibid.;* see Chapter 10, p. 146.
10. Edward T. Hall, *The Hidden Dimension,* Doubleday and Company, Inc., Garden City, N.Y., 1966. Also published as a paperback (Premier Books, New York, 1969), p. 1.
11. *Ibid.,* p. 116.
12. Michael Argyle and Janet Dean, "Eye-contact, Distance and Affiliation," *Sociometry,* 28, No. 3 (September, 1965), 289-304. See also Michael Argyle, *The Psychology of Interpersonal Behavior,* Penguin Books, London, 1967.
13. Mardi J. Horowitz, Donald F. Duff, and Lois O. Stratton, "Personal Space and the Body-Buffer Zone," *Archives of General Psychiatry,* 2 (December, 1964), 651-656. Also reprinted in Harold M. Proshansky, William H. Ittelson, and Leanne G. Rivlin, *Environmental Psychology: Man and His Setting,* Holt, Rinehart and Winston, Inc., 1970, pp. 214-220.
14. O. Michael Watson and Theodore D. Graves, "Quantitative Research in Proxemic Behavior," *American Anthropologist,* 68, No. 4 (August, 1966), 971-985, and O. Michael Watson, *Proxemic Behavior,* Mouton, The Hague, 1970.
15. Miriam Liebman, "The Effects of Sex and Race Norms on Personal Space," *Environment and Behavior,* 2, No. 2 (September, 1970), 208-246.

16. Robert Sommer, *Personal Space: The Behavioral Basis of Design,* Prentice-Hall Inc., Englewood Cliffs, N.J., 1969, Chapter 3.
17. Alan Lipman, "Chairs as Territories," *New Society,* 9, No. 283 (April 20, 1967), 564-566.
18. *Ibid.,* p. 566.
19. Irwin Altman and William H. Haythorn, "The Ecology of Isolated Groups," *Behavioral Science,* 12 (1967), 169-182; reprinted in Proshansky, Ittelson, and Rivlin, *Environmental Psychology: Man and His Setting,* pp. 226-239.
20. *Ibid.,* p. 228.
21. *Ibid.,* p. 230.
22. Irwin Altman, "Ecological Aspects of Interpersonal Functioning," in *Behavior and Environment: The Use of Space by Animals and Men,* ed. Aristide H. Esser, Plenum Press, New York, 1971, p. 297.
23. Sim Van Der Ryn and Murray Silverstein, "The Room, A Student's Personal Environment," in *People and Buildings,* ed. Robert Gutman, Basic Books, New York, 1972, pp. 370-383.
24. Sommer, *Personal Space,* p. 78.
25. Bernard Steinzor, "The Spatial Factor in Face to Face Discussion Groups," *Journal of Abnormal and Social Psychology,* 45, No. 3 (July, 1950), 552-555.
26. *Ibid.,* p. 552.
27. F. L. Strodtbeck and L. J. Hook, "The Social Dimensions of a Twelve-Man Jury Table," *Sociometry,* 24 (1961), 397-415.
28. Sommer, *Personal Space,* p. 20.
29. For a more detailed examination of this relationship, see Alton J. De Long, "Dominance-Territorial Relations in a Small Group," *Environment and Behavior,* 2, No. 2 (September, 1970), 170-191.
30. See, for example, Harold J. Leavitt, "Some Effects of Certain Communication Patterns on Group Performance," *Journal of Abnormal and Social Psychology,* 46 (1951), 38-50; reprinted in Maccoby, Newcomb, and Hartley, *Readings in Social Psychology,* Holt, Rinehart and Winston, Inc., New York, 1958, pp. 546-563.
31. Sommer, *Personal Space,* Chapter 7.
32. Bruno Bettelheim, *The Informed Heart,* Macmillan, New York, 1960, p. 207.
33. Paul Goodman, "Seating Arrangements: An Elementary Lecture in Functional Planning," in *Utopian Essays and Practical Proposals,* Random House, Inc., Alfred A. Knopf, Inc., New York, 1964, pp. 156-181.
34. Charles Winick and Herbert Holt, "Seating Position as Nonverbal Communication in Group Analysis," *Psychiatry,* 24 (May, 1961), 171-182.
35. Robert Sommer, *Design Awareness,* Rinehart Press, San Francisco, 1972.

Three

Architectural space

"We shape our buildings and afterwards our buildings shape us."
Winston S. Churchill[1]

Perhaps the most obvious way we have intervened in nature to create an entirely new environment is with our buildings. Architecture could be considered an instrument whose central function is to modify the environment in our favor. Each building has the function of protecting us from the natural environment so that we can conserve more of our energy for productive work. Without buildings we would always be engaged in direct interaction with the elements. Within buildings we live an encapsulated existence, secure and protected from wind, rain, sun, snow, noise, and other people. Thus buildings represent the second line of defense for human beings, with clothing as the first. This is well illustrated in Figure 3.1, taken from a discussion by James M. Fitch.[2]

Fitch goes on to note that the building is not a totally impermeable container. Rather, its outer surfaces could be compared with the permeable membranes that protect the developing child within the mother's body, selecting and rejecting various impinging environmental forces. While it is likely that the womb is the optimum environment for the child's growth, the building is not always optimum for our purposes. Buildings are more than a container to shield us from various outside influences; in addition, their internal arrangements and facilities should support people and en-

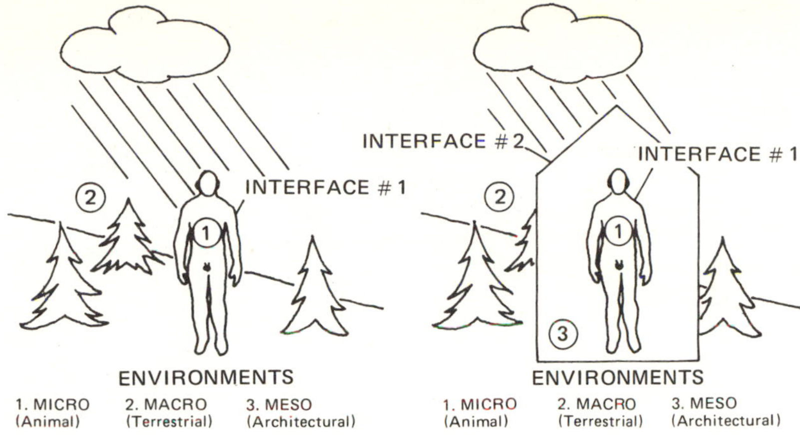

Fig. 3.1 Buildings as our second line of defense. (James M. Fitch, "Experimental Bases for Aesthetic Decisions," *Environmental Psychology: Man and His Physical Setting*, The New York Academy of Sciences, *Annals* Vol. 128, 1965. Adapted by permission.)

hance their ability to perform the functions for which buildings are designed.

In earlier times buildings were not so finely differentiated within. According to Aries, rooms had no fixed functions in European houses until the eighteenth century.[3] Many activities took place in the same room without much specialization of areas according to use. Today, as more and more of our activities take place inside and as society becomes more complex, the range of specialized activities within buildings requires more careful consideration in design. Great advances in building technology now provide us with the necessary flexibility to design for human needs and purposes, but we still lack accurate knowledge about people's perceptions and behaviors in the built environment.

The building as a system containing systems of rooms

In contrast to Chapter 2, which dealt with people in a room environment, this chapter deals with the larger spaces of building interiors. The interactions of occupants and their immediate environment, as in a room, was seen to operate as a system. This small-room system may be regarded as a subsystem of the larger system contained within a building. Although the people-environment system seems to operate as a unit at the room level, it is not entirely independent of the larger system within which it is embedded. This is illustrated by John Zeisel's discussion of how the relationship between different rooms can be altered by changes in the floor plan to accommodate different behavior patterns.[4]

Zeisel contrasts the use of living rooms, dining rooms, and kitchens in white, middle-class families, in working-class, New York Puerto Rican families, and in moderate-income, southern black families. He shows how the physical design can support the underlying social needs of those who use the environment. In so doing, Zeisel emphasizes the importance of discovering the latent as well as the manifest function of spaces and their behavior. In its obvious or "manifest function," a kitchen would be the place to cook. However, in addition, there may be many "latent functions" or social meanings—for example, the kitchen as a place where women prove that they do their job in society well. Usually these underlying social meanings are not consciously known to the inhabitants of a setting, but they may be discovered by observation of behavior. Such observations enabled Zeisel to draw up floor plans that would best accommodate the behaviors of the three groups observed.

For the middle-class white families, according to Zeisel, the living room is the place where guests are entertained. The central social concern is how well the guests mix and get along. An efficient, compact kitchen close to the living room makes it easy for the wife to join the social occasion. "Since the living room is the symbol of sociability, the main entrance into the apartment can easily open directly into the living room, symbolizing the hospitality of the family."[5] A plan for this group is shown in Figure 3.2(a).

Zeisel discovered different social meanings in the homes of working-class New York Puerto Rican families. Here meals are eaten in the kitchen. "The more time the lady of the house spends there the better mother she is, in the eyes of her friends and family."[6] Since the mother spends so much of her time in the kitchen, it is important that she have access to the front door, to keep track of other family members' comings and goings. On the other hand, Zeisel writes, the living room holds a special meaning for the family. On the walls are often "pictures of John F. Kennedy, large plaster saints and the high school diplomas of the children . . . . The living room containing all of these political, religious, and economic icons of the family is, for them, a revered space, to relax away from the kitchen and for special occasions."[7] It is best placed far from the entrance, as in the plan shown in Figure 3.2(b).

Among middle-class southern blacks in South Carolina, Zeisel found that whenever possible the families would have a separate dining room.

> The need for a dining area apart from the the kitchen reflects the culture of blacks in the south where traditional spicy and smelly foods like chitterlings, spare ribs, and fried chicken were cooked. Whereas originally a separate dining room was needed to get away from the cooking smells, even as cooking habits changed, the social importance of having a separate dining area remained.[8]

For this group the plan includes a separate dining room, as seen in Figure 3.2(c).

It seems likely that by altering the floor plans in the direction of the social

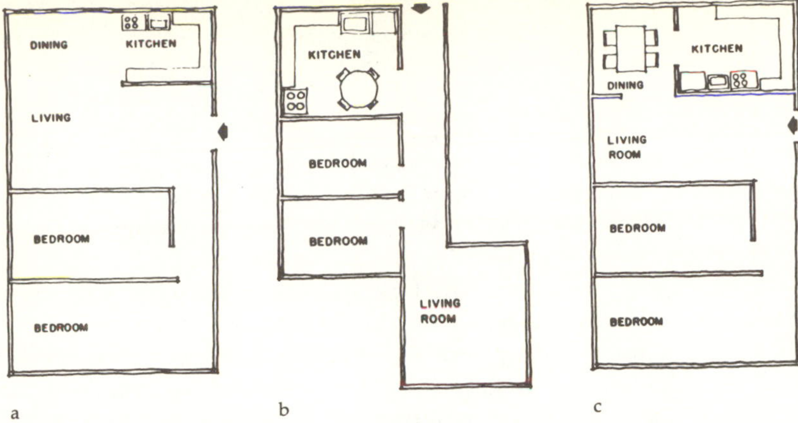

Fig. 3.2 Apartment floor plans for three different groups: (a) white middle class, (b) New York Puerto Rican working class, and (c) moderate-income southern black. (John Zeisel, "Fundamental Values in Planning with the Nonpaying Client," in *Architecture for Human Behavior,* ed. Charles Burnett, *et al.*, Philadelphia Chapter of the American Institute of Architects, 1971, pp. 27-28. Reprinted by permission.)

meanings for the residents the desired behavior patterns can more easily be realized. The special social functions of each room may be more effectively separated from those of other spaces and the quality of life in the entire apartment improved. Unfortunately, we do not have the data to prove or disprove such hypotheses. They await the construction of the apartments and the measurement of the behavior changes. But it is clear that there are linkages between the discussion of personal space and room geography in Chapter 2 and architectural space, considered here. Colbert expressed the idea in these terms:

> Any unit must be judged in comparison to the units next smaller and below, or next larger and above, and ultimately by comparison of the world itself. A common example of this thought is the design of a room. To design a room adequately, we must first judge its contents (furniture, for example), and then the whole of the house is related to its immediate environment and upward through ever enlarging cycles of concern to the universe itself.[9]

Anthropozemic and anthropophilic buildings

Izumi suggests classifying buildings in terms of the importance of psychosocial dimensions.[10] He uses the terms "anthropozemic" (foreign to man) and "anthropophilic" (attractive to man). Anthropozemic buildings would be those, such as power plants, designed primarily to contain objects of machinery, equipment, and other inanimate things. At the other end of the

scale would be buildings used solely by human beings, such as nursing homes, penitentiaries, and psychiatric hospitals (Figure 3.3). In between would be buildings that contain objects and human beings in varying proportion. The more anthropophilic the building, the greater the importance of psychosocial considerations in design. The hospital, as a very anthropophilic building, has been selected for detailed consideration in this chapter. Although the hospital is distinct in form and function, the design problem may be expected to have parallels in other types of buildings. In considering hospitals we will discuss several examples of studies linking human perception and behavior to architectural designs. These types of studies may well be applied in other architectural contexts.

Physical aspects of the hospital environment

Some of the earliest and most thoughtful attempts to relate human needs to architectural design have been made in studies of hospital design. It is not surprising that hospitals should provide pioneer studies in this field, for:

> . . . the symbiotic relationship between the architectural container and the men and processes contained is nowhere clearer than in the modern hospital. Here we find every degree of biological stress, including that of birth and death.[11]

Within a hospital seriously ill patients will traverse the full spectrum of experience in their perception of environment. In the operating room, under anesthesia, the patient will not be conscious of the environment, but as the patient gradually recovers during the period of convalescence, response to the surrounding environment will become of greater importance. Color, lights, sounds, and smells can become active elements of

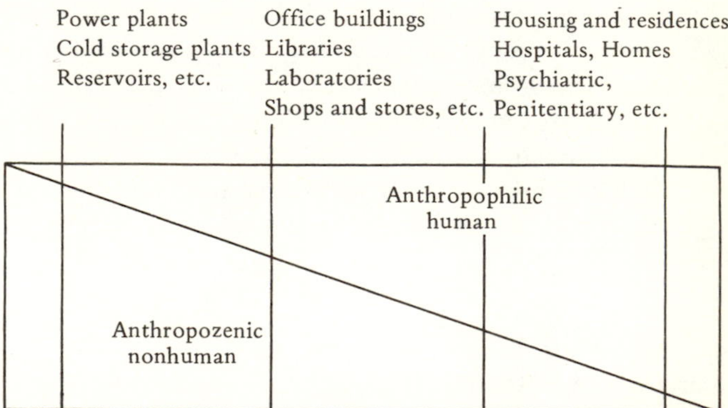

Fig. 3.3 Anthropozemic and anthrophilic buildings. (Kiyoshi Izumi, "Psychosocial Phenomena and Building Design," *Building Research,* 1965, II, 9. Adapted by permission.)

therapy. All such factors should be analyzed objectively to remove all unnecessary stress, which could delay convalescence.

For the doctors and staff, special conditions should be created to enhance their performance where this is not in conflict with the needs of the patient. In the operating room, for example, where the visual acuity of the surgeon is of prime importance, the color of the walls, uniforms, and towels and the lighting fixtures all are geared to aiding the physician to see clearly. On the other hand, in the same room the patient and operating staff can have opposing requirements, which must be resolved. The patient requires warm, moist air, while the staff, under nervous tension, should ideally have dry, cool air. In this case the more pressing needs of the patient take precedence, so "the staff sweats, suffers and recovers later."[12]

Within hospitals are found some of the most extreme examples of special environments created and controlled for specific purposes. Some examples of these are:

> . . . the hyperbaric chamber where barometric pressure and oxygen content are manipulated in the treatment of both circulatory disorders and gas gangrene; the metabolic surgery suites where body temperatures are reduced to slow the metabolic rate before difficult surgery; the use of saturated atmospheres for serious cases of burn; artificially-cooled, dry air to lighten the thermal stress on cardiac cases; the use of electrostatic precipitation and ultraviolet radiation to produce completely sterile atmospheres for difficult respiratory ailments or to prevent cross-infection from contagious diseases. Here the building is not merely manipulating the natural environment in the patient's favor but actually creating totally new environments with no precedent in nature as specific instruments of therapy.[13]

The creation of special environments in hospitals emphasizes the effect of physical factors such as light, temperature, sounds, and smells on the well-being of the hospital patients and personnel. It parallels the kind of research noted in the second chapter on people-machine systems in the careful fitting of people to the physical environment surrounding them. Precise measurements of variations in specific physical elements such as light level can be matched with human responses in laboratory-type settings. Much work in the field of building research is of this type. Here we will be more concerned with social and psychological criteria for hospital design, which are more difficult to measure and currently less amenable to precise design solutions.

Varying criteria for hospital design

There are many different criteria with which to assess the effectiveness of hospital design. Hospitals have an organizational structure that includes several different decision-making groups. These can be described in terms

of three distinct levels: technical, managerial, and institutional.[14] In the hospital situation the technical level would include nursing, medical, and paramedical groups; the managerial level, administrative and executive officers; and the institutional level, the boards of directors and trustees. Each of these groups has its own function, which would of course affect the group perceptions of what good hospital design should include.

The three main kinds of functions are: (1) utility, providing health care services to the community; (2) amenity, satisfying individual or personal requirements of people working or staying in the hospital; and (3) expression, the symbolic function of the hospital facility as a public institution and a haven for the ill. The technical staff is most concerned with the utility function; the managerial staff, with the amenity; while the requirements for expression are most directly in the province of the institutional level. Clearly, these are not entirely separable, but the classification scheme does at least illustrate the way very different criteria might be developed for assessing the design of a hospital. Conflicts can occur as well. Summer noted that the grounds are often landscaped to provide an impressive vista for visitors or state legislators (expressive function) instead of for the use of patients (utility function).[15] Here we will be most concerned with the utility function, which, as the studies below indicate, is closely intertwined with the amenity function and has symbolic or expressive elements as well.

For a number of reasons hospital design is not generally based on users' needs. We have indicated that there are several different decision-making groups, all with their own major concerns, which may not coincide with those of the patients. Another problem is that the user is not generally the paying client, so the architect may not be strongly motivated to think in terms of user needs.[16] An article by Lindheim on factors that influence hospital design notes the strong hold of habitual patterns.[17] These rather than modern medical requirements often determine design. In using the example of a radiology department, Lindheim indicates that most planning guides for such departments are made by x-ray equipment companies, whose advice is influenced by the needs of their equipment rather than the needs of the radiology department or the total hospital.[18] Thus the influence of commercial firms, the varying perspectives of different decision-makers, and the perpetuation of past spatial patterns may lead the architect to ignore the needs of the principal-user groups. Research on how hospitals are actually used may help to achieve a balance.

Experimental testing of nursing-unit designs

With the aid of a federal grant a 570-bed hospital building was constructed in Rochester, Minnesota, for use as a laboratory for research into various aspects of hospital architecture.[19] The first major research project con-

ducted in the hospital investigated the impact of three different nursing-unit designs on the activities and subjective feelings of nursing personnel. The study is discussed here because it represents a sample of the type of testing within specially designed buildings that must be undertaken to test theories relating architectural design to behavior. It will become apparent in the discussion that follows that the costs on a national basis for such experimental buildings could easily be made up in improved efficiency of staff and resources if better designs were adopted. The argument might be made to justify the development of experimental cities[20] or plans on a regional level, though we will not resume the argument in the chapters that follow.

The literature on nursing-unit design contains three basic floor plans. These three designs are known as the "radial" (or round), the "single-corridor," and the "double-corridor" (or racetrack). They are illustrated in Figures 3.4, 3.5, and 3.6. Proponents of each design claim advantages in improved efficiency of nursing and care for patients. To test the differences, nursing units of each type were constructed in the hospital, standardized as much as possible except for the difference in basic design. All the testing was completed within one hospital with the same administration and hospital staff. This greatly strengthened the experimental controls, as many of the extraneous variations could thus be minimized. Four

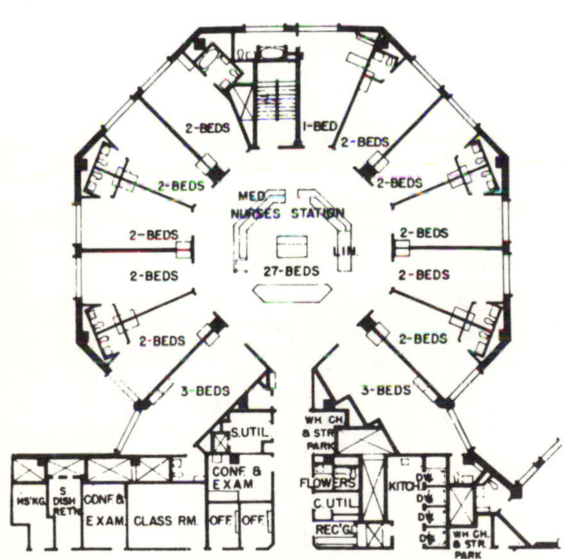

Fig. 3.4 Radial nursing unit. (From "Influence of Nursing-Unit Design on Activities and Subjective Feelings of Nursing Personnel," by David K. Trites *et al.* Reprinted from *Environment and Behavior*, Vol. 2, No. 3 (December, 1970), pp. 304-306 by permission of the publisher, Sage Publications, Inc.)

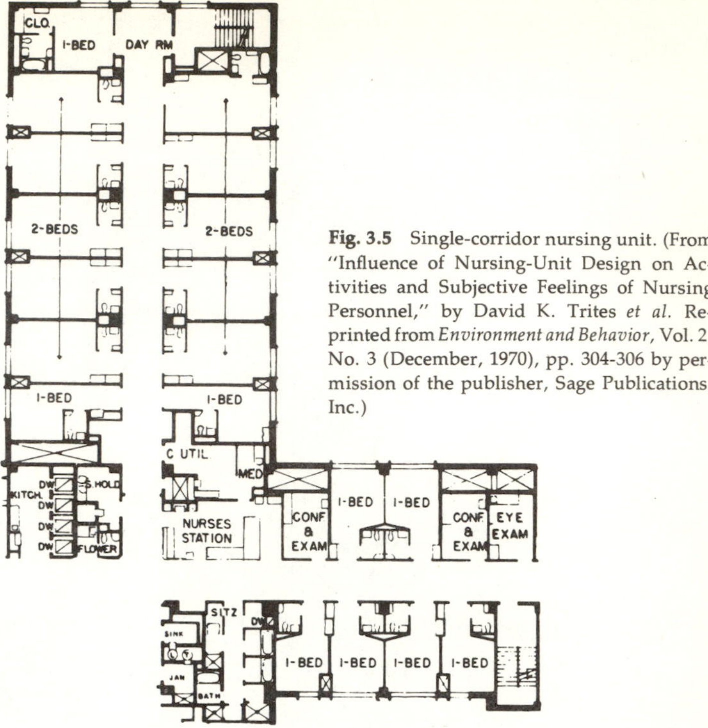

Fig. 3.5 Single-corridor nursing unit. (From "Influence of Nursing-Unit Design on Activities and Subjective Feelings of Nursing Personnel," by David K. Trites *et al.* Reprinted from *Environment and Behavior,* Vol. 2, No. 3 (December, 1970), pp. 304-306 by permission of the publisher, Sage Publications, Inc.)

nursing units of each of the three types were examined on the day, evening, and night shifts by work-sampling and staff questionnaires. A total of 590 different people participated over a period of 82 days.[21]

The results of the study indicated that in most instances the radial design was superior to the double-corridor. The double-corridor in turn was superior to the single-corridor design. There was overwhelming preference for the radial units among the staff. This was apparent in the answers to questionnaires and in a supporting study of absenteeism and accidents to staff members. There were significantly fewer absences among personnel assigned to the radial units, and relatively fewer accidents as well. The data collected indicate that staff on radial units spent more time with their patients and less time in travel than the staff on the single- and double-corridor units. The staff on the radial units could devote more time to the patients and still have time left over. Without any decrement in nursing attention, the number of patients on the radial wards could be increased. Small wonder that physicians and patients also preferred the radials and physicians felt that the radial design enhanced the quality of patient care.

An interesting aspect of the study was the discussion of time saved on the radials expressed in dollars. On the basis of average salaries for each class of personnel in 1968, it was calculated that the extra travel costs in the

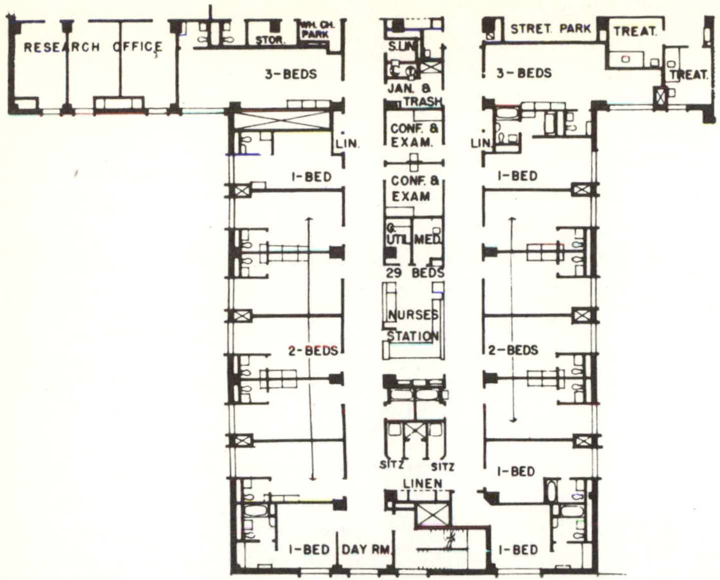

Fig. 3.6 Double-corridor nursing unit. (From "Influence of Nursing-Unit Design on Activities and Subjective Feelings of Nursing Personnel," by David K. Trites *et al.* Reprinted from *Environment and Behavior*, Vol. 2, No. 3 (December, 1970), pp. 304-306 by permission of the publisher, Sage Publications, Inc.)

single- and double-corridor units compared to the radials amounted to approximately $77 per bed per year.[22] Furthermore, the extra time spent with the patients on the radial units was worth $67 per bed per year compared with the double-corridor units and $97 per bed per year compared with the single-corridor units. Clearly, in this case good design is not only preferred by the personnel and patients, it also pays. It is not always so easy to express savings in human suffering in terms of money, but the discussion of mental hospital design that follows indicates that here as well the monetary savings could be considerable.

Function as a basis for mental hospital design

An ideal approach to mental hospital design based on the specific needs of the patients has been outlined by Humphry Osmund.[23] His paper provides a clear presentation of the type of information required at the beginning of the architectural design process and emphasizes the need for interdisciplinary cooperation—for example, in this case the psychiatrist must provide the information used by the architect in the final design. Since the ap-

proach is one that should have broad applicability in a variety of types of buildings, it will be discussed in some detail.

Function is considered the basis for psychiatric ward design. The architect's task is to design a building that fulfills the functional requirements expressed by the client. In the case of the mental hospital, the psychiatrist would be in the best position to present the needs of the patients. In other cases, of course, the behavior of those most involved in the use of the building should be observed, and they should be consulted as to their perceptions and preferences. Osmund starts with the needs of patients, which he examines in some detail, and then sets up a series of rules derived from these needs. These rules provide the architect with explicit goals that should be built into the design.

The needs of mentally ill people have been more clearly defined than those of most other groups. In a most general sense, these people have in common a reduced capacity to relate to others. This disturbance in interpersonal relationships is usually caused by changes in (1) perception, (2) feeling, and (3) thinking.

Disturbances in perception among schizophrenic people, the largest group in North American mental hospitals, range from illusions to distortions and hallucinations of a gross nature. Distortions may occur in visual, auditory, tactile, or olfactory perception as well as perception of time or one's own body. Changes in spatial perception are often accompanied by changes in perception of the body. This uncertainty about the integrity of the self is likely to be aggravated by huge corridors and enlarged spaces so often found in mental hospitals.[24] Susceptibility to auditory hallucinations among the mentally ill makes it necessary to avoid disturbing echoes and other auditory peculiarities in the design. Similarly, hypersensitivity to odors makes it more important than in general hospitals to have a sufficient number of well-ventilated and properly functioning toilets.

Changes in mood among the mentally ill make interaction difficult. In large groups, mood swings can reverberate through many people, leading to panic and social disintegration. But stabilization of mood changes is more likely in a small group of people who know and understand each other. Privacy and a place of their own are essential to avoid being overwhelmed by emotions of their companions.

Changes in thinking may make even the simplest actions difficult. The ward design should therefore make the patient's biological life easy by having washing and toilet facilities close by and easily located. Where memory failure is a factor, rooms as well as lockers and drawers, must be clearly labeled. Ample provision of mirrors may be helpful if the patients can cover them when they want to.

After listing the needs of mental patients, Osmund goes on to derive a set of rules based on the needs. These rules translated into architectural forms should then serve the needs of the patients. The rules are as follows: Patients must be neither overcrowded nor overconcentrated. There must

be provision for a path of retreat to a private place. Opportunities to form beneficial relationships must be provided, and psychosocial needs met, as, for example, through the supportive relationships developed among groups of four to eight people. It is necessary to reduce ambiguity and uncertainty while preserving and limiting choice. The preservation of personality and individuality must be enhanced, as, for example, by avoiding design emphasizing mass living and herd existence and providing possibilities for individuality in personal living space. Sexual segregation should be avoided, so that the patients can maintain the necessary social skills required by our culture.

With a list of needs and a set of rules derived from them the architect is in a much better position to design the building, even though the exact design solution for each rule has not been specified. This degree of precision must await further study, experimentation with forms, and analysis of new physical arrangements. However, at least some of the rough outlines are possible, and the architect has clear guiding principles to work with. In a general sense, he knows that for a mental hospital "sociopetal" design is desirable. "Sociopetal design encourages, fosters and even enforces the development of stable interpersonal relationships such as are found in small face-to-face groups." This type of design is opposite to "sociofugal design" (both terms coined by Osmund), which "prevents or discourages the formation of stable human relationships."[25] Examples of buildings that are generally sociofugal are railway stations, hotels, mental hospitals, and jails, while the tepee, igloo, and Zulu kraal could be considered highly sociopetal. Furniture arrangements that encourage interaction would be sociopetal, while those that impede interaction would be sociofugal.

Behavioral mapping

To analyze the behavior of patients on the mental ward or in any other building, the technique of *behavioral mapping* could be useful. This technique, developed by Ittelson, Rivlin, and Proshansky, provides a means of relating various types of behavior to their physical locus.[26]

Behavioral mapping is concerned with various categories of behavior, physical locations, and a technique for relating the one to the other. It begins with the architect's floor plan. This is a scale drawing of a building indicating all the physical areas. The labeling of the rooms shows the types of behavior expected in each—e.g., "bedroom," "dining room," or "bathroom." However, the architect's floor plan implies very broad groupings of behavior, and, in addition, each room could be subdivided into smaller, more precise activity areas. For example, the living room could be separated into areas for TV viewing, reading, and conversation. The types of behavior to be considered are determined on the basis of observed behavior. The categories for analysis must be precise and relevant to the

particular problem. They are usually the most common types of behavior that occur in the areas under consideration. In the psychiatric ward they might be, for example, eating, sleeping, reading, talking, or watching television. A table is constructed with the physical areas as rows and the categories of behavior as columns. Then observers take samples at various times of the day, noting which types of behavior occurred at which locations. Thus a quantitative summary of the use of various portions of the building for various purposes is obtained. This can be analyzed by comparing the actual and expected use of spaces, the use by different categories of people and at different times of the day or week.

An application of the behavioral mapping technique to determine the appropriate bedroom size in a psychiatric ward was made by Ittelson, Proshansky, and Rivlin.[27] The study was conducted in the psychiatric wards of three large metropolitan hospitals, which they referred to as "private," "state," and "city." In each ward time-sample observations of the activities at various locations and who participated were noted on a special recording form. Later the observed behavior was grouped into three major categories: "*isolated passive* (lying in bed, asleep or awake, and sitting alone); *isolated active* (personal hygiene, reading, arts and crafts, and other miscellaneous activities); *social* (patient-patient, patient-staff, and patient-visitor interactions)."

The results are revealing: Although the most common type of behavior in all wards is isolated-passive, there are consistent differences depending on the number of beds per room. In all hospitals, as the number of patients in the room increases, isolated-passive behavior comprises an increasingly larger proportion of the total activity in the bedroom. Social behavior correspondingly decreases. Contrary to much of the folklore of the psychiatric ward, it is not the private room but the large, multiple-occupancy room that provokes patients' withdrawal. Apparently patients in private rooms perceive themselves as having a wider range of choice in behavior. They feel free to select from several options and do choose more equally among the possibilities. On the other hand, patients in the larger multiple-occupancy room seem to see their range of choice as severely limited. They are more likely to engage in isolated-passive behavior than anything else, and while they are in the room they may spend from two thirds to three quarters of their time lying on their beds, either asleep or awake. Most of the withdrawn behavior in the multiple-occupancy rooms takes place in the presence of others. While this was true for the hospitals observed, it should be noted that single hospital rooms are seen as undesirable by many other groups.[28]

The study described above was part of a much larger program of studies utilizing the same technique. Behavior in corridors and public rooms was recorded as well. In addition, separate mappings at two different times were completed to test the stability of patterns and to assess the effect of certain physical changes on patient activities.

The experimenters introduced certain physical changes in the ward to attempt to influence behavior. In the first set of observations in the city hospital, they noticed that a solarium at the far end of a corridor was seldom used, probably because of its location and its inadequate and uncomfortable facilities. The researchers decided to study the effect on the use of the solarium of providing comfortable, attractive, and carefully laid out seating arrangements. The changes were made, and the second behavioral mapping took place after a time judged sufficient to allow enduring patterns of use to be established. As expected, the solarium became more popular with the patients, and its share of the total activity in all the public areas almost doubled. The proportion of activities in the other dayroom and the corridor dropped. But this was not all.

What occurred was not merely an increase in the solarium's share of behavior; in addition, there was a marked change in the entire pattern of activities throughout the public areas of the ward. The entire ward showed an increase in the proportion of active behavior. Interestingly, the type of behavior within the solarium showed the smallest shift, with a slightly larger proportion of active behavior. The dayroom showed a shift toward more isolated-passive behavior, while in the corridor there was an opposite effect. Here isolated-passive behavior dropped sharply and active behavior rose. The precise amounts of change in types of behavior are interesting, but more important is the clear demonstration of the way a physical change in one area of the ward affected the distribution and type of behavior in all the other areas. One should think of the ward as a social system in which a change in one element has ramifications in all the other parts.[29] The implications of this for the design of buildings and rooms within buildings are immense, and yet we know practically nothing about the exact effects or the process. Here, as at all the other levels of environment considered in this book, a much more sensitive appraisal of the linkages between human actions and the results, both planned and unplanned, must be undertaken.

Broad implications for the design of buildings resulted from the studies utilizing the behavioral mapping technique carried out by faculty from the environmental psychology program of the City University of New York. Their work indicated that patterns of behavior within particular wards were remarkably stable over time. This might be expected in other buildings as well. Daily, weekly, or seasonal cycles would show up regularly as explainable alterations to the expected pattern. The study group demonstrated the utility of the behavioral mapping technique, which could be used to test these generalizations and the effects of alterations in design. Their dictum, "people, policies, and partitions clearly do affect patterns of behavior,"[30] suggests that there are several ways to bring about change. Wise planning would involve weighing each method before deciding which was most appropriate for the particular situation. Improvements in design based on systematic observations could well lead to better be-

havioral patterns. However, there is no assurance that a good design will always work by itself, for any potential benefits could quickly be nullified by an unenlightened policy decision. Good design, then, is important but not all-powerful.

Dominance and territoriality in a mental hospital ward

A fascinating study, which owes its inspiration directly to animal studies on territoriality, was undertaken by Aristide H. Esser and his associates.[31] It has been well established by ethologists that territorial factors are crucial in maintaining stable social and spatial arrangements in many, or even most, animal populations.[32] The "pecking order," for example, in a flock of chickens, involves a ranking of all the members of the group so that the most dominant member has the first peck at the food and first choice of where it wants to go. It serves the group purposes for the acceptance of the hierarchical order, reduces fighting and other forms of social tensions, and thus provides stability, which is beneficial for all. The pecking order, however, does not imply a total dominance. Each animal, regardless of its position in the dominance hierarchy, is dominant in its own territory. This has often elicited comparisons with the general success of the home team in sports.[33] But how far it is valid to carry comparisons of animal and human behavior is a question that only further research can answer. Altman has indicated the directions such research should follow in his analysis of the concept of territorial behavior in humans.[34] The study by Esser and his associates in a psychiatric ward is an excellent example of the type of research required.

The dominance and territoriality of twenty-two schizophrenic patients in a psychiatric ward was studied by two main types of observation. The first, "location observation," corresponds closely to behavioral mapping in that, on a time-sample basis (observations made every half hour), it notes the locations of patients and their behavior in the ward. The second method involves recording the interactional behavior of the individual patients. As long as a patient was on the day ward, observations were recorded noting when and for how long he interacted with specific patients or staff members and also who initiated the interaction. The combined data resulting from location observation and the interactional behavior enabled the researchers to rank the patients in terms of dominance and also to determine where each patient kept himself. An example of the preprinted plans on which the data were recorded is seen in Figure 3.7.

Examination of the data revealed that only half the patients made use of all the available space on the day ward. Others had certain areas or rooms they avoided. About half the patients spent much of their time in one particular area, which was defined as their territory according to the proportion of time they spent there. The range of movement in the ward

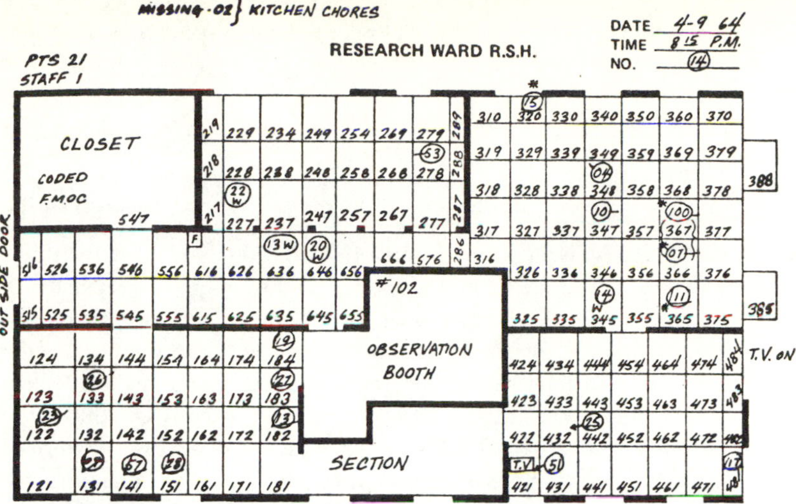

Fig. 3.7 Plans for ascertaining territories on ward. (Aristide H. Esser *et al.*, "Territoriality of Patients on a Research Ward," in *Recent Advances in Biological Psychiatry*, VII. ed. J. Wortis, Plenum Publishing Corporation, New York, 1965, p. 40. Reprinted by permission.)

and the establishment of territory appeared to be related to the position in the dominance hierarchy. Patients in the upper third of the hierarchy were free to move wherever they wanted and did not seem to need to establish ownership of a particular place. Those in the middle third of the hierarchy showed some restrictions in range and seemed to establish territories located in positions of maximum traffic flow, where contacts were most likely. In contrast, patients in the bottom third of the hierarchy were moderately restricted in range and seemed to seek out secluded spots in which to withdraw from the likelihood of contacts. In one extreme case the patient was seen in one spot in 273 out of a total 330 observations.[35]

To explore more fully the relationships between territory and dominance, the researchers made counts of the aggressive behavior of each patient as noted in the clinical records. It turned out that certain patients were never mentioned, others were occasionally involved in fights, while a final group regularly displayed aggressive behavior. These data show that aggressive behavior is related to a person's instability in the dominance hierarchy and to possession of a territory. Recently admitted patients tend to fight more often than do older inhabitants of the ward. Those patients shifting from active-drug to placebo medication or vice versa also exhibited instability in the dominance hierarchy and a greater likelihood of involvement in incidents of aggression. Protection of territory also seems involved, for, of all the patients who possessed territory, only three did not show aggressive behavior.

The parallels with findings from studies of animal behavior are clear, and the implications for physical design and future research are fully explored by Esser and his associates, in a later book based on a symposium on the use of space by animals and people.[36] Once pinpointed, areas of friction due to undesirable territorial patterns may be modified by manipulating the physical structure of the ward, as, for example, in the case cited above when the solarium was changed. If it is confirmed unequivocally that schizophrenics are "personal-space sensitive," it may prove possible to use data on the use of space in the clinical estimate of a patient's mental state. Horowitz provides many examples of what might be done in citing from studies of a large military psychiatric ward.[37] Paluck and Esser indicate that, among severely retarded men, changes in territorial behavior may be used as an indicator of changes in clinical condition.[38] In all these examples the importance of individual differences in human spatial behavior is evident.

Medical center design in the total urban context

Before leaving the discussion of hospital facilities it might be pertinent at this point to go to a slightly different scale to consider medical center design in the context of the total urban setting.[39] Building in the old familiar pattern has become less and less satisfactory, due to unprecedented advances in biomedical sciences. There is a trend toward viewing the hospital as part of the larger community, with doctors and patients passing in and out. The mental hospital has an open door and serves as a major mental health resource for the community rather than as a "final dumping ground" or "museum of pathology."[40] This involves a shift from an imposition of health care services on the basis of the providers' ideas and concepts (mainly bed care today) to a more participatory system involving a wide range of types and sites of care from the home to the intensive care unit. It also involves a much greater preventive emphasis and a growth of new decentralized units in multiuse clinics, community centers, housing projects, schools, places of work, mobile units, and the home itself.[41] All these changes will require new approaches by the physical planner or designer.

Hermann H. Field illustrates such an all-encompassing planning approach in describing the Tufts-New England Medical Center Project.[42] Such planning and design must include concern for the social and behavioral as well as the physical, so that an interdisciplinary team is desirable. It is essentially an ecological approach toward the artificial environment. While one aspect of the planning is aimed outward from the medical center, involving the surrounding community, the other aspect is directed inward to the growth of the complex itself. The usual hospital complex serves as a barrier to the free flow of the neighborhood or appears as an

alien body in its midst. The mere size and scale of such complexes may be intimidating.

The Tufts-New England Medical Center Project adopted the strategy of seeking to mesh the complex with its surroundings at all edges, to become a part of the community. It was decided to let the normal neighborhood life flow under the medical facilities by creating shopping plazas and pedestrian walkways connecting with the new station of the transit system. Air rights over the streets were obtained from the city and pedestrian loops within the medical complex were integrated with the streets below (Figure 3.8). Instead of acting as a barrier, the medical center could thus serve as the southern gateway to the city's center between the residential and industrial structures in the South End and the city's retail core (Figure 3.9).

Summary

The studies cited in this chapter, dealing with the topic of how we perceive and use architectural space, have all been selected to focus on one type of organization, the hospital. The complexity of the interactions that take

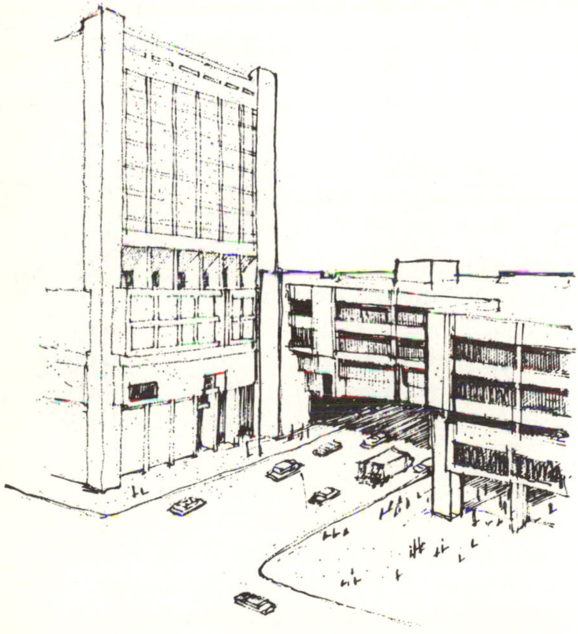

Fig. 3.8 Streets and roads flowing under medical complex. (Hermann H. Field, "Medical Center Planning and Design Within the Total Urban Setting." *World Hospitals,* April 4, 1970, pp. 4-8. Reprinted by permission.)

Fig. 3.9 Medical complex as gateway to city center (Hermann H. Field, "Environmental Design Implications of a Changing Health Care System," in *Environment and Cognition*, ed. William H. Ittelson, Academic Press, Inc., New York, 1973, p. 138. Reprinted by permission.)

place between people and their environment at this scale is evident. In the case of the hospital, there is no simple answer for the architect because people on the technical, managerial, and institutional levels have differing perspectives on what is important in hospital design. The research described did not canvass all these levels equally. The main concern was with the perceptions and behavior of the patients and the technical-level staff. But even so, there are clearly different perspectives for the designer to include in his deliberations. The needs of the doctors, nurses, and patients may not always coincide exactly, and yet all should be taken into account.

Though it is desirable to use the behavior and perceptions of the main users as a major source of information in the design process, this has not been the general practice: Habit may lead to a replication of past hospital designs; radiology departments may be designed to suit the equipment rather than the needs of the radiologists or their patients; and the main users are often not the paying clients. These factors have a familiar ring, for they also operate at the scale of the room.

There are many other parallels with the findings of the studies discussed in Chapter 2. Again we see the application of a wide range of techniques, including tightly controlled experiments, systematic observations of behavior, interviewing, and experimental alteration of the environment. The

precision possible in measuring physical elements of the environmental interaction contrasted with the difficulty of defining or measuring the human dimensions. The importance of cultural factors was once more emphasized in Zeisel's descriptions of behavioral differences in the use of apartments. The topic of territoriality and dominance as well as the role of individual differences was again stressed in the work of Esser and his associates on territoriality in a mental hospital ward. The importance of conceiving of the human-environment interaction as a system was seen in the study of Ittelson, Proshansky, and Rivlin. They observed that modifications in the physical arrangements of one room induced changes in the type and amount of behavior throughout a psychiatric ward. But again, the process by which such effects occur remains a mystery.

The built environment was the main focus of attention, and its influence was seen as important but not all-powerful. The framework of Constance Perin may be useful here. She suggests that what we tend to describe as "the effect of environment on behavior is actually the extent to which the environment responds to the stimulus of human demands, . . . the physical resources for behavior can be undermining to supportive, absent to present, rare to ubiquitous." We could conceive of an "environmental response continuum" ranging from "structured, directive, and authoritarian" to "minimal, open, and flexible."[43] At either end of the continuum, there may be high adaptive costs. At one end there is too much structure, while at the other there may be too much choice. In the center of the continuum would be the *congruent environmental response*, which would be a plan based on the behavior of the future inhabitants, a plan that would enhance a sense of competence in everyday behavior.

Although the examples in this chapter centered on the hospital, it is assumed that the same kinds of effects would be seen in other types of architectural space.[44] Stable patterns of behavior in any building could be measured by the behavioral mapping technique. New buildings could be designed to suit the inhabitants or intelligently altered or improved by using Osmund's general guidelines.

Many problems of how people perceive and use architectural space cannot be approached as directly as in the case of nurses and their preferences for ward design. For this reason it is necessary to develop systematic frameworks for observation such as behavioral mapping. In investigating a segment of the real world, some selection of what to attend to is imperative. For the mental hospital certain commonly recurring types of behavior were selected. For other buildings other behaviors might be more important, and these could be observed and located. But the basic technique is transferable and could be usefully applied in a variety of settings at the scale of architectural space.

Similarly, one can imagine transferring a framework such as Osmund's to develop guidelines for the design of factories, playgrounds, or prisons, simply by starting with the needs of the space user and deriving from these

a set of rules for the designer. Such guidelines do not provide specific design elements for the solution but at least they indicate the kinds of solutions that should be sought. A concrete example criticizing a playground design in terms of the degree to which it satisfies a set of essential needs is provided by Constance Perin[45] (Table 3.1). Such a criticism will alert the designer to flaws that might otherwise be overlooked. The particular playground analyzed is seen to be especially deficient in its relationship to the surrounding environment and in provision of opportunities for self-expression.

The interlocking of different levels of environment was stressed in this

Table 3.1 Table on Essential Striving Sentiments and Relative Resources

Essential striving sentiment	Relevant resource
Physical security (safety from cars on parking lot; dogs; wandering and getting lost; protection from one's own immaturity)	4-foot high cyclone wire fence (relates positively)
Sexual satisfaction (play materials and forms that offer tactile and motor activities related to body awareness; sexual identity; and pleasurable muscular feelings)	Slide, swings (relates positively)
Expression of hostility (shouting, competitive play, digging, destroying, taking apart objects)	Playspace abuts building (relates negatively)
Expression of love (cooperative play)	Individual-use play equipment; waiting for turns (relates negatively)
Securing of love Securing of recognition (gaining adult approval)	Entire site relates negatively; playspace offers little opportunity for children to create and take pride in games or building of forms that can bring self-recognition
Expression of spontaneity (thinking up new games)	Fixed play elements whose use is limited (relates negatively)
Orientation of place in society and places of others (looking around now and then)	Site is tucked behind the building instead of sharing in the active passers-by movement
Securing and maintaining membership in a definite human group	Site is discontinuous with life around it (relates negatively)
Sense of belonging to a moral order	The siting of the building's parking lot, which gets better sun all day, has shade trees and a view toward a pleasant residential street, symbolizes all too well the priority for society of car storage over playspace

SOURCE: Reprinted from *With Man in Mind* by Constance Perin by permission of The M.I.T. Press, Cambridge, Mass., p. 126. Copyright © 1970 by Constance Perin. Reprinted by permission.

chapter. Zeisel's study showed the importance of relating the behavior in kitchens and living rooms to the total apartment layout. It was seen that changes in one room, the solarium, led to behavioral modifications throughout a mental ward. Failure to relate the playground to the surrounding neighborhood activities was criticized. Two examples stressed the importance of rooms as subsystems within the larger system of a building. The other emphasized viewing the design in terms of the larger segment of surrounding environment. A final example by Field also illustrated how medical center design can help or hinder social relations in the broader community. It is to the broader community that we now turn as we move up in scale to the level of the small town and neighborhood.

Notes

1. According to William Michelson in *Man in His Urban Environment: A Sociological Approach* (Addison-Wesley Publishing Co., Reading, Mass., 1970, p. 168), there is an unwritten rule that anyone writing on the social influence of architecture must quote Winston Churchill's statement on the opening of the House of Commons after its wartime destruction. To avoid any criticism on this score, I have included it at the start. Another unwritten rule of the same type applying to the design of hospitals is to quote Florence Nightingale, who conceived the first requirement of the hospital "to do no harm to the patient." In both cases, of course, the quotes are apt, to the point, and worth citing. See also the quote of J. K. Wright on geosophy in Chapter 6 and those at the beginning of the other chapters.
2. James Marston Fitch, "Experiential Bases for Aesthetic Decisions," *Annals of the New York Academy of Sciences,* 128 (1965), 706-714, reprinted in Harold M. Proshansky, William H. Ittelson, and Leanne G. Rivlin, *Environmental Psychology: Man and His Physical Setting,* Holt, Rinehart and Winston, Inc., N.Y., 1970, pp. 76-84. For further discussion see James Marston Fitch, *American Building: The Environmental Forces That Shape It,* Houghton Mifflin Company, Boston, 1972.
3. Philippe Aries, *Centuries of Childhood,* Alfred A. Knopf, Inc., New York, 1962, as quoted in Hall, *The Hidden Dimension,* p. 104.
4. John Zeisel, "Fundamental Values in Planning with the Nonpaying Client," in *Architecture for Human Behavior,* ed. Charles Burnette, *et al.*, Philadelphia Chapter of the American Institute of Architects, Philadelphia, 1971, pp. 23-30.
5. *Ibid.*, p. 26.
6. *Ibid.*, p. 26.
7. *Ibid.*, p. 26.
8. *Ibid.*, p. 27.
9. Charles Colbert, "Naked Utility and Visual Chorea," in *Who Designs America,* Princeton Studies in American Civilization No. 6, ed. Laurence B. Holland, Doubleday & Company, Inc., Anchor Books, Garden City, N.Y., 1966, pp. 214-235.
10. Kiyoshi Izumi, "Psychosocial Phenomena and Building Design," *Building Research,* No. 4 (July-August, 1965), 9-11, reprinted in Proshansky *et al.*, *Environmental Psychology,* pp. 569-573.
11. James Marston Ftich, "Experiential Bases for Aesthetic Decision," (see footnote 2, p. 82.

12. *Ibid.*, p. 83.
13. *Ibid.*, p. 83.
14. J. J. Souder, W. E. Clark, J. I. Elkind, and M. B. Brown, "A Conceptual Framework for Hospital Planning," in Proshansky *et al.*, *Environmental Psychology*, pp. 579-587.
15. Sommer, *Personal Space*, p. 90.
16. Zeisel, "Fundamental Values in Planning with the Non-paying Client," in *Architecture for Human Behavior*, ed. Charles Burnette *et al.*
17. Roslyn Lindheim, "Factors Which Determine Hospital Design," *American Journal of Public Health*, 56, No. 10 (1956) 1668-1675, reprinted in Proshansky *et al.*, *Environmental Psychology*, pp. 573-579.
18. *Ibid.*, p. 577.
19. David K. Trites, Franklin D. Galbraith, Jr., Madelyne Sturdavant, and John F. Leckwart, "Influence of Nursing-Unit Design on the Activities and Subjective Feelings of Nursing Personnel," *Environment and Behavior*, 2, No. 3 (December, 1970), 303-334.
20. Athelstan Spilhaus, "The Experimental City," *Daedalus*, 96, No. 4 (Fall, 1967), 1129-1146.
21. Trites *et al.*, "Influence of Nursing-Unit Design on the Activities and Subjective Feelings of Nursing Personnel," p. 319.
22. *Ibid.*, p. 331.
23. Humphry Osmund, "Function as the Basis of Psychiatric Ward Design," *Mental Hospitals* (Architectural Supplement), 8 (April, 1957), 23-30, reprinted in Proshansky *et al.*, pp. 560-569; see also Osmund, "Some Psychiatric Aspects of Design," in *Who Designs America*, Princeton Studies in American Civilization No. 6, ed. Laurence B. Holland, Anchor Books, Garden City, N.Y., 1966, pp. 281-318.
24. For a discussion of architectural features that are suitable for the mentally ill, see K. Izumi, "Some Architectural Considerations in the Design of Facilities for the Care and Treatment of the Mentally Ill," paper prepared for the American Schizophrenia Foundation, second revised draft, April 28, 1967.
25. Osmund, "Function as the Basis of Psychiatric Ward Design," p. 567.
26. William H. Ittelson, Leanne G. Rivlin, and Harold M. Proshansky, "The Use of Behavioral Mapping in Environmental Psychology," in Proshansky *et al.*, *Environmental Psychology*, pp. 658-668.
27. William H. Ittelson, Harold M. Proshansky, and Leanne G. Rivlin, "The Environmental Psychology of the Psychiatric Ward," in Proshansky *et al.*, *Environmental Psychology*, pp. 419-439.
28. Sommer, *Personal Space*, p. 87.
29. Sommer found a similar serendipitous outcome of ward changes in the elderly people's home described in Chapter 2. See Sommer, *Personal Space*, p. 86.
30. Ittelson, Proshansky, and Rivlin, "The Environmental Psychology of the Psychiatric Ward," p. 438.
31. Aristide H. Esser, Amparo S. Chamberlain, Eliot D. Chapple, and Nathan S. Clein, "Territoriality of Patients on a Research Ward," in Proshansky *et al.*, *Environmental Psychology*, pp. 208-214; originally printed in *Recent Advances in Biological Psychiatry*, ed. J. Wortis, Plenum Publishing Corporation, New York, 1965.
32. For a well-written popular overview, see Robert Ardrey, *The Territorial Imperative*, Dell Publishing Company, New York, 1966.
33. See comments on this by Sommer, *Personal Space*, p. 14.
34. Irwin Altman, "Territorial Behavior in Humans: An Analysis of the Concept," in *Spatial Behavior of Older People*, ed. Leon A. Pastalan and Daniel H. Carson, The University of Michigan-Wayne State University, Institute of Gerontology,

Ann Arbor, Mich., 1970, pp. 1-24.

35. Esser *et al.*, "Territoriality of Patients on a Research Ward," p. 214.
36. Aristide H. Esser, ed. *Behavior and Environment: The Use of Space by Animals and Men*, Plenum Press, New York, 1971.
37. Mardi J. Horowitz, "Human Spatial Behavior," *American Journal of Psychotherapy*, 19, No. 1 (1965), 20-28.
38. Robert J. Paluck and Aristide H. Esser, "Territorial Behavior as an Indicator of Changes in Clinical Behavioral Condition of Severely Retarded Boys," *American Journal of Mental Deficiency*, 76, No. 3 (1971), 284-290.
39. Hermann H. Field, "Medical Center Planning and Design Within the Total Urban Setting," *World Hospitals*, 6, No. 2 (April, 1970), 80-89.
40. Leonard J. Duhl, "The Changing Face of Mental Health," in *The Urban Condition*, ed. Leonard J. Duhl, Simon and Schuster Clarion Books, New York, 1963, pp. 59-75, quotations on p. 66.
41. See Footnotes 28 and 40.
42. Hermann H. Field,"Environmental Design Implications of a Changing Health Care System," in *Environment and Cognition*, ed. William H. Ittelson, Academic Press, New York, 1973, pp. 127-156.
43. Constance Perin, *With Man in Mind*, The MIT Press, Cambridge, Mass., 1970, p. 42.
44. For some idea of the range of architectural spaces that have been studied from the perspective of human perception and behavior, see *Architectural Psychology*, Proceedings of the Conference at the University of Strathclyde, ed. David V. Canter, RIBA Publications, London, 1970; Proshansky, Ittelson, and Rivlin, eds., *Environmental Psychology;* Robert Gutman, ed., *People and Buildings*, Basic Books, New York, 1972; and the Proceedings of the annual Environmental Design Research Association Conferences.
45. Perin, *With Man in Mind*, p. 126.

Four

Small towns and neighborhoods

"People conform in a high degree to the standing patterns of the behavior settings they inhabit."
Roger Barker

Small towns and neighborhoods are the next step up in scale from architectural space. They represent units with very different qualities from those of the largely interior spaces of buildings. It is no longer the internal arrangement of space that is important, but instead the external spatial arrangements of many buildings and public spaces—that is, broad-scale site planning. Small towns and neighborhoods may coincide in size, but often there are great qualitative differences between them, especially if one includes densely settled urban neighborhoods. However, they are often linked together because of the intimate type of human contact both are presumed to provide.[1] In contrast with the larger unit of the city, which provides the possibility of specialized contacts, the neighborhood or small town provides, or should provide, the warmth of everyday human contacts in the regular round of work and play. Although it could be argued that the small town differs from the neighborhood, they will both be considered here because of the similarity in scale and the intimate nature of contact. The really significant change occurs when one steps up to the larger size and more impersonal atmosphere of the city. Even within cities, however, the urban neighborhood to some degree retains the characteristics of smaller size and the possibility of more personal contacts.

The discussion carries on from where it left off last chapter—that is, on architectural aspects of site planning. We will briefly sketch examples of some of the broad links between site planning and social behavior. The question of spatial determinism is raised here, as it is at the base of many suggestions for the design of small towns and neighborhoods.

Site planning and social behavior

Robert Gutman's short review of site planning and social behavior notes that the dominant research direction has been conceived of as the influence the site plans exert through their regulation of the communication process.[2] However, he also cites three other pathways through which site plans can be linked with social behavior. The earliest of these was housing studies conducted by social reformers around the turn of the present century. They emphasized the provision of adequate amenities, particularly fresh air, heat, illumination, and sanitation, and demonstrated the high rates of morbidity and mortality associated with slums and tenement housing.[3] Another rarely applied approach is that of the influence of site plan aesthetics. The third approach concerns the way objects acquire symbolic meaning. Here we will consider briefly the influence of the site plan on the communication process. This may be kept in mind as a contributing factor in the creation of neighborhoods.

It is often assumed that the physical features of the site plan establish a specific network through which messages are exchanged. In many respects this is similar to the previous discussion on communication as a function of seating position. The site plan blocks off certain avenues of contact between persons, while others are open, even emphasized. The presence of barriers or open paths in specific places influences the probability of contact, which could lead to communication. Such factors have been examined in studies of the intensity of social relations among residents in various types of housing projects. The role of propinquity—that is, the distance between dwelling units—has also been stressed in the promotion of social relations. William H. Whyte, Jr., for example, states that "in suburbia friendship has become almost predictable,"[4] and:

> Given a few physical clues about the area, you can come close to determining what could be called its flow of "social traffic," and once you have determined this you may come up with an unsettlingly accurate diagnosis of who is in the gang and who isn't.[5]

Figures 4.1 and 4.2, taken from his book, *The Organization Man,* illustrate how location in a particular portion of the site plan may determine who is likely to get together with whom and the clique to which one belongs.

Once formed, the social patterns may persist in spite of a constant

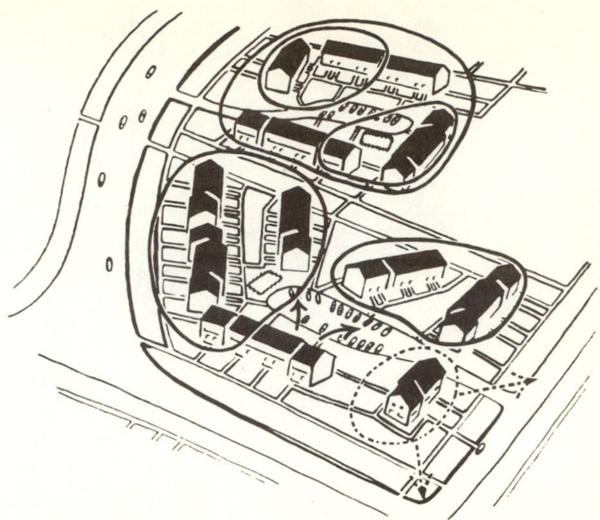

Fig. 4.1 What makes a court clique. (William H. Whyte, Jr., *The Organization Man*, Simon & Schuster, Inc., New York, 1956, pp. 344-345. Copyright © 1956 by William H. Whyte, Jr. Reprinted by permission of Simon & Schuster, Inc.)

turnover of residents. Thus Whyte found that each court tended to produce its own pattern of behavior, and "whether newcomers become civic leaders or bridge fans or churchgoers will be determined to a large extent by the gang to which chance has now joined them."[6] The persistence of particular behavior patterns in specific settings described by Whyte has remarkable resemblances to the behavior setting theory of ecological psychology, which will be discussed later in this chapter.

The initial physical factors that produced the friendship patterns noted by Whyte are such things as the placement of a stoop or the direction of a street, the placement of play areas selected by children, adjacent driveways, or adjoining lawns. Others have noted the importance of the placement of doors or windows and, in dormitory buildings, the locations of lavatories.[7]

In a thoughtful review of the literature, William Michelson assesses the degree to which such physical determinism is valid.[8] He points out that proximity becomes a factor in friendship under two conditions. The first is homogeneity, or perceived homogeneity; the second is the need for mutual aid. Both of these were present in the samples of suburbia studied by Whyte. All the people were very similar in age, income, and social class and had similar problems requiring mutual aid during the settling-in period. In public housing, where the residents perceive themselves as different from their neighbors, there is a decided lack of neighboring. One excep-

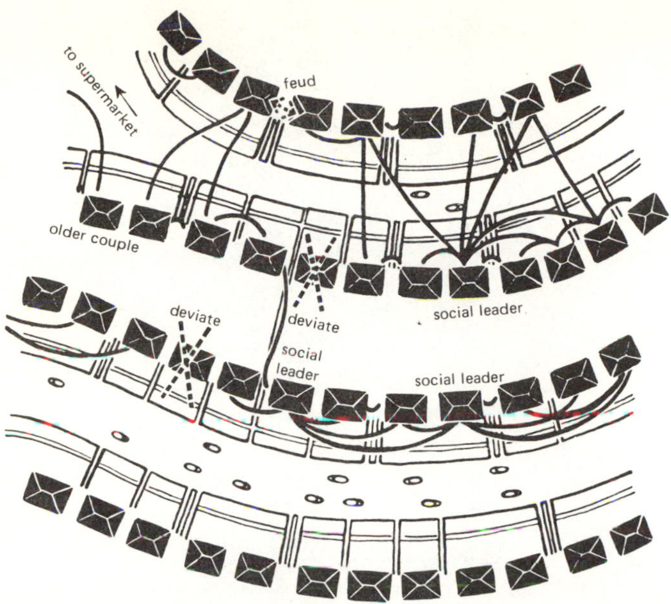

Fig. 4.2 How homeowners get together. (William H. Whyte, Jr., *The Organization Man,* Simon & Schuster, Inc., New York, 1956, pp. 344-345. Copyright © 1956 by William H. Whyte, Jr. Reprinted by permission of Simon & Schuster, Inc.)

tion, which has been noted, is the case of husbandless mothers, to whom neighboring within public housing is important. Here is an obvious case of the need for mutual aid.

A recent study by Robert Athanasiou and Gary Yoshioka provides further evidence to weigh in the controversy regarding physical determinism.[9] Their carefully designed study in an Ann Arbor suburb lent some support to the contention that homogeneity or perceived homogeneity is an important factor in friendship formation. They found, for example, that propinquity did not overcome differences in stage of life cycle. On the other hand, they concluded that "propinquity plays a part in the formation and maintenance of friendships among women who may have little in common besides life-cycle stage."[10] Furthermore, the strength of propinquity is such that it may cause people who expect high levels of interaction to misperceive each other favorably. This was supported by the interesting observation of Athanasiou and Yoshioka that:

> Subjects' perceptions of their friends' political beliefs were better able to discriminate between friends and non-friends than were actual political affiliation. It seems likely that the spatially closer the friend, the more an individual would tend to distort her perceptions in a direction which would be compatible with maintaining contact through propinquity.[11]

So although homogeneity was seen as important, it was at times a misperceived homogeneity, covering up differences that otherwise might prove damaging to friendship formation.

Athanasiou and Yoshioka found that the spatial factor was exceedingly important in explaining the distribution of friendships. Although initially inclined to consider homogeneity as more important than propinquity in determining the intensity of social relationships, they rejected this hypothesis in light of their data. Women in their study area did have a higher proportion of their more intense friendships close by. The correspondence between the spacing of dwelling units (D. U.) and the distribution of intense friendships was so close the at the authors suggest that, "it may well be that the distribution of friendships can be manipulated by the spacing of the D. U."[12] To test this contention, further studies in different types of site plans and with dwelling units differently spaced are called for. It seems likely that some of the contradictions in findings thus far could be explained by major differences in housing types and site plans. Further evidence of differences in behavior associated with variations in site plan are provided in Oscar Newman's book *Defensible Space*.[13]

Defensible space

According to Oscar Newman, the crime problems facing urban America result from a breakdown of the social mechanisms that once kept crime in check. "The small-town environments, rural or urban, which once framed and enforced their own moral codes, have virtually disappeared. We have become strangers sharing the largest collective habitats in human history."[14] The anonymity, isolation, irresponsibility, and lack of identity with our surroundings in large cities have provided an atmosphere in which criminal activities flourish. Many large housing projects built in haste and without much thought or reference to tradition have become prime targets for criminal attacks. Newman contends that this need not be so. He believes that it is possible, by manipulation of building and spatial configurations, to create areas for which people will adopt concern. This is referred to as "defensible space."

> Defensible space is a model for residential environments which inhibits crime by creating the physical expression of a social fabric that defends itself. All the different elements which combine to make a defensible space have a common goal—an environment in which latent territoriality and sense of community in the inhabitants can be translated into responsibility for ensuring a safe, productive, and well-maintained living space. The potential criminal perceives such a space as controlled by its residents, leaving him an intruder easily recognized and dealt with.[15]

Newman's conclusions are based on an analysis of the nature, pattern, and location of crime in urban residential areas in the United States. In housing units for all income groups, the inhabitants, project managers, and police were interviewed. In addition, data on crime, vandalism, and maintenance costs were gathered. The burden of the argument is borne by analysis of the extensive and detailed statistics gathered by the New York City Housing Authority, which is responsible for 150,000 units of public housing (19 percent of all United States public housing). These statistics pinpoint not only the nature of the crime, the victim, and the offender, but also the specific buildings and interior locations in which the crime took place. This has enabled Newman to compare crime rates with specific design features, as well as with general characteristics such as project size and building heights.[16]

Table 4.1 illustrates the relationship between project size, building heights, and crime rates in a sample of 100 New York City projects. It is clear that higher crime rates are associated with higher buildings. The highest crime rates are found in the very large projects with higher buildings. But large project size by itself is not always associated with higher crime rates. It is apparently possible to maintain high densities without higher crime rates, providing the building height remains low. This is dramatically demonstrated by Newman's use of the method called a "comparison of coupled projects."

The comparison of coupled projects method involves statistical comparison of two housing projects in which all the variables that could affect the crime rate are identical except the projects' physical design characteristics. The Brownsville and Van Dyke projects in New York City provide the possibility of such a comparison. They are located adjacent to each other in Brooklyn. The projects are strikingly different in physical design while they house comparatively identical populations in size and social charac-

Table 4.1 Project Size and Building Height versus Crime

Project size	Building height	
	Equal to or less than 6 stories	Greater than 6 stories
Equal to or less than 1000 units	N = 8 M = 47 SD = 25	N = 47 M = 51 SD = 23
Greater than 1000 units	N = 11 M = 45 SD = 26	N = 34 M = 67 SD = 24

NOTE: N = number of cases; M = mean number of crimes per thousand; SD = standard deviation.
SOURCE: Reprinted with permission of Macmillan Publishing Co., Inc. from *Defensible Space* by Oscar Newman p. 28 Copyright © 1972 Oscar Newman. First published in Great Britain in 1973 by The Architectural Press, London.

teristics. The statistics speak for themselves. As Table 4.2 indicates, there were many more criminal incidents in the Van Dyke project that consists mainly of fourteen-story, high-rise slabs with large open spaces between them (Figure 4.3). In contrast, the Brownsville project consists mainly of low, walk-up and elevator buildings three to six stories high. The buildings and their arrangement divide the site into smaller, more manageable zones. All the residents and police interviewed perceive the Brownsville project as smaller and more stable than the adjoining Van Dyke project. Intruders feel more cautious about invading the privacy of residents at Brownsville than at Van Dyke, where they wander with impunity in the interior corridors. Newman notes that Brownsville is far from perfect in terms of defensible space design, but it clearly offers the residents a better opportunity to identify with specific small spaces, which they come to feel responsible for and maintain surveillance over.

The importance of citizen surveillance in the streets was previously emphasized by Jane Jacobs, who also recognized the need for the maintenance of human scale in urban neighborhoods and decried huge impersonal projects.[17] But Newman carries the ideas much further, and the high quality of the crime statistics available enables him to pinpoint the most vulnerable areas within buildings. Figure 4.4 shows where felonies were committed within public housing projects in New York City. The mean felony rate goes up with the size of the building. It is in the interior public spaces of the largest buildings, removed from natural surveillance, that a high proportion of the crimes take place. The elevator, a prime example of an area removed from surveillance, is the single most vulnerable area. Much of the danger can be removed by design features with built-in surveillance, such as lobbies clearly visible to the street, windows facing halls, and features that foster relations between a limited number of people sharing a common portion of the project area.

Newman's work is an excellent example of the type of research that analyzes the behavioral results of site plans after the buildings have been constructed. Through the use of aggregate data on crime and housing design, he has indicated a host of questions that should be investigated

Table 4.2 Comparison of Crime Incidents

Crime incidents	Van Dyke	Brownsville
Total incidents	1,189	790
Total felonies, misdemeanors, and offenses	432	264
Number of robberies	92	24
Number of malicious mischief	52	28

SOURCE: New York City Housing Authority Police Records, 1968. Reprinted with permission of Macmillan Publishing Co., Inc., from *Defensible Space* by Oscar Newman, p. 47. Copyright © 1972 by Oscar Newman. First published in Great Britain in 1973 by The Architectural Press, London.

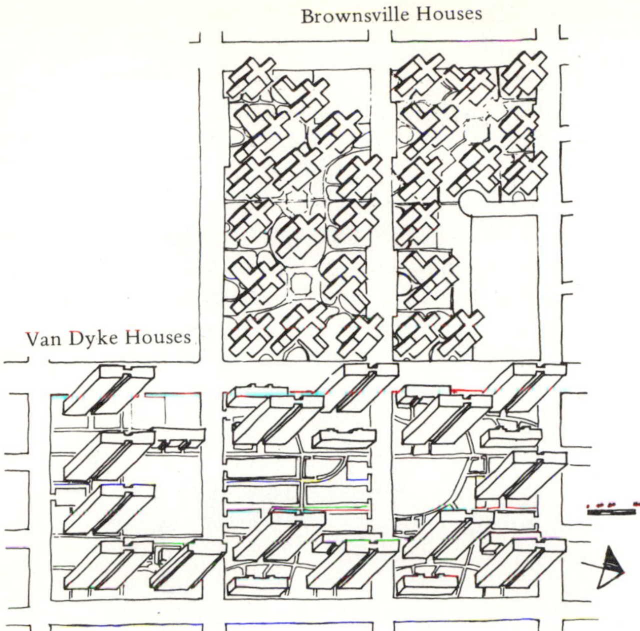

Fig. 4.3 Site plan of Brownsville and Van Dyke houses. (Reprinted with permission of Macmillan Publishing Co., Inc. from *Defensible Space* by Oscar Newman. Copyright © 1972 by Oscar Newman.)

further. His emphasis on the need for a sense of community in urban neighborhoods and the role of site planning in fostering it underlie much of the concern in this chapter. He is not espousing a simplistic physical determinism but advocating a type of design that provides the residents with better possibilities of relating to and controlling their own neighborhood environment. Later in this chapter we will return to a discussion of neighborhoods as perceived by people living in them, but first we will turn to a consideration of the small town, which has often served as a model for the type of intimate human contacts ideally provided by neighborhoods.

Small-town atmosphere

Small towns have not generally been the center of attention in studies of people-environment interaction. However, there have been a few notable exceptions, such as the work of J. B. Jackson and *Landscape* Magazine, in which many articles have appeared that evoke and explain some of the characteristic features of town and small-city atmosphere.[18] Another exception is the output of the research laboratory of ecological psychology that was founded in the small town labeled Midwest, Kansas. The two contrast-

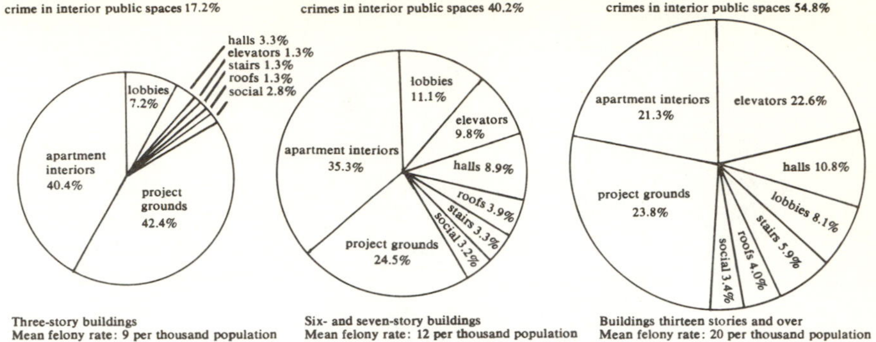

Fig. 4.4 The place of occurrence of crimes in buildings of different heights. (Reprinted with permission of Macmillan Publishing Co., Inc. from *Defensible Space* by Oscar Newman. Copyright © 1972 by Oscar Newman.)

ing approaches could be considered as examples of the types of studies available. These range from the anecdotal observations of a casual but perceptive participant to the theoretical framework developed by trained, systematic observers.

The pages of *Landscape* contain many excellent articles on the towns and smaller cities of America. Probably the best of these were written by J. B. Jackson, the founder of the magazine and its editor for seventeen years. His articles illustrate clearly the aim of the magazine, reflected in the title *Landscape*—to provide a forum for informed accounts of various facets of the humanized landscape; in other words, to explore the relationship between the people and the land. To do this in depth, one must understand not only our relation to the world but to one another. This kind of understanding is evident in J. B. Jackson's inimitable articles, which convey to the reader not only the authentic atmosphere of places described by a sensitive participant-observer but an appreciation of ways in which the landscape reflects past and present notions of beauty and utility. Criticism of orthodox planning is combined with specific suggestions as to local sources of vitality that should be understood, supported, and enhanced by thoughtful planning. Jackson's discerning eye has scanned the full range of landscape elements, but here only a couple of essays on smaller cities or towns will be sampled.

One such center is Optimo City, really not one place but:

> . . .a hundred or more towns, all very much alike, scattered across the United States from the Alleghenies to the Pacific, most numerous west of the Mississippi and south of the Platte. When, for instance, you travel through Texas and Oklahoma and New Mexico and even parts of Kansas and Missouri, Optimo City is the blur of filling stations and motels you occasionally pass; the solitary traffic light, the glimpse up a side street of an elephantine courthouse surrounded by elms and

> sycamores the brief congestion of mud-spattered pickup trucks that slows you down before you hit the open road once more.[19]

Optimo City is an old-fashioned sort of place, with its courthouse in the center and streets leading out from each side of the square. The ugly old courthouse serves both as a center for civic activity and as a symbol for civic pride—unlike many modern civic centers, in which the "civic consciousness has been divorced from everyday life, put in a special zone all by itself."[20] All the local offices are there. The agricultural agent, the district nurse, and the Boy Scouts, the Red Cross, and the federal soil conservation service are present. But the courthouse and its square interrupt the flow of traffic in all four directions, so that "a sluggish eddy of vehicles and pedestrians is the result."[21] Beyond is a regular grid pattern of streets, where "you are likely to see a tractor turn into someone's drive with wisps of freshly cut alfalfa clinging to the vertical sickle bar."[22] Here the ties between town and country have not yet been broken.

> Saturday, of course, is the best time for seeing the full tide of human existence in Sheridan County. The rows of parked pickups are like cattle in a feed lot; the sidewalks in front of Slymaker's Mercantile, the Ranch Cafe, Sears, the drugstore, resound to the mincing steps of cowboy boots; farmers and ranchers, thumbs in their pants pockets, gather in groups to lament the drought (there is always a drought) and those men in Washington, while their wives go from store to movie house to store.[23]

But all is not well in Optimo City. The chamber of commerce points out that the town is not growing and that it has to depend on "a few hundred tightfisted farmers and ranchers for its livelihood."[24] They suggest tearing down the courthouse to provide much-needed parking space, widening Main Street into a four-lane highway, and attracting new businesses catering to tourists. Jackson states that these suggestions are sensible on the whole, but he seems dubious of their wisdom as he comments further:

> Translate them into more general terms and what they amount to is this: if we want to get ahead, the best thing to do is break with our own past, become as independent as possible of our immediate environment and at the same time become almost completely dependent for our well-being on some remote outside resource.[25]

He adds that the formula has been successful for many American towns but argues that Optimo City should perhaps have a different destiny, one that remains associated with the surrounding landscape of which it has been a part for the past century or so. By retaining ties with the local landscape, the residents may also better retain their independence. Jackson's essay reflects a keen awareness of the frame of mind that may induce small-town planners, imbued with conventional planning wisdom, to destroy some of

the few remaining sources of local character and vitality in a perhaps illusory attempt to attract new outside sources of revenue. A similar situation is described in a more recent article on small towns by Pierce Lewis.[26]

Lewis studies American small towns on two levels. First he considers them as a group, and then he examines the small town of Bellefonte, Pennsylvania, as a case study. He contends that their importance is far greater than their current population numbers would indicate, for small-town imagery and mythology are central to the American experience. Americans have loved small towns, which is reflected in American literature, replete with major novels having small-town settings. The writers are not alone in their predilection for small towns. Rather, they are reflecting a national consensus. Much of the sentiment associated with small towns "may represent simply the lavender-scented nostalgia which surrounds the dimly remembered past."[27] The negative aspects, such as the narrow provincialism, may be forgotten as the former town dweller looks back from his new perspective in the problem-ridden cities:

> What remains is a belief that small towns are built at a scale which permits the inhabitants to exercise meaningful control over their local environment if they choose to do so—that community relations are more stable and neighborly than in a city—that the cost of urban living may be too high if it is bought with physical danger and hatred.[28]

In the future small towns may seem even more desirable as images of their neighborliness and stable community relations contrast with the feeling of loss of control in larger, more anonymous, and even dangerous urban areas. Changes in technology may also enable small towns to play a new role in contemporary society. However, in spite of change in taste and technology, Lewis foresees difficulties that may prevent small towns from adapting constructively to changing circumstances.

In the case of Bellefonte, three generations of economic stagnation and loss of population have caused great physical and psychological damage. The atmosphere of despair and pessimism has created a grim conservatism that leads to a reluctance to entertain new ideas and a clinging to established patterns of thought and behavior. Yet, in a desperate effort to induce growth, the town has narrowed its options still further, as highway building and urban renewal have led to demolition of distinguished old buildings, which have been replaced by gas stations and outlets for chain stores or restaurants (Figures 4.5 and 4.6). Unfortunately, it is mainly outsiders and returned natives who see the possibilities of retention and restoration of the earlier architectural atmosphere to create a physically attractive town and to shore up the ebbing community spirit. Most residents are sure it will not work.

Both Jackson and Lewis see the small town as an area in transition. The danger is that, in the effort to solve economic problems, the townspeople may overlook the real advantages of the local environment and destroy the

Fig. 4.5 Georgian mansion built 1838 destroyed by the Atlantic Refining Company, 1958, to make room for a gas station. (Peirce Lewis, "Small Town in Pennsylvania." Reproduced by permission from the *Annals of the Association of American Geographers,* Vol. 62, 1972, p. 340.)

very vitality they hope to create. The demoralization resulting from economic stagnation may lead to a search for stereotypic outside solutions rather than an analysis of the local sources of strength and stability. Effective planning requires a clear understanding of the relationship between behavior and environment. This can be arrived at only through an assessment of how the area in question is perceived and used by the local inhabitants. Jackson displays such an empathetic understanding in an article on another aspect of small town life.

Highway strips

In "Other-directed Houses," J. B. Jackson describes not an older pattern that could be obliterated through a lack of awareness but a newer, developing pattern that is still not really understood, the highway strip development.[29] These have often been labeled as "longitudinal slums" or "blighted areas" and described with some accuracy as ugly and confusing. But blanket condemnation and suppression, as called for in certain highway reform programs, may be a bit premature. First we have to understand them and their functions. Jackson distinguishes two main groups of establishments: those such as factories, warehouses, truck depots, service stations, used-car lots, and shopping centers, which serve the working economy; and restaurants, cafés, nightclubs, amusement parks, drive-in movies, souvenir stands, and motels, which serve our leisure. He contends that more and more of our leisure time is spent cruising for pleasure along these highway strips, a role that Main Street, the park, or the courthouse

Fig. 4.6 Gas station that replaced the mansion in Bellefonte, Pennsylvania. (Peirce Lewis, "Small Town in Pennsylvania." Reproduced by permission from the *Annals of the Association of American Geographers,* Vol. 62, 1972, p. 340.)

square used to play in the free time of our pedestrian predecessors. Here too is all the vitality of the mixed public, "teenagers, transients, people in search of amusement, doing business, alone and in groups."[30] The strip businesses develop "other-directed architecture" to attract the passing public. Since the people are going by rapidly in cars, the new architectural style is designed to attract attention by flashy, conspicuous façades, exotic decoration or landscaping, and a lavish use of lights, colors, and signs. Jackson characterizes it as "a kind of folk art in midtwentieth century garb,"[31] and describes the strip as:

> . . .a jumbled reminder of all current enthusiasms—atomic energy, space travel, Acapulco, folksinging, computers, Danish contemporary, health foods, hot rod racing and so on.[32]

Part of its appeal is based on the informal atmosphere, but more important is the fact that it does not try to separate the automobile from its driver. Jackson suggests that planners, architects, industrial designers, and advertising experts work together to introduce some order and harmony. This could perhaps be accomplished by grouping at frequent intervals some of the leisure-time activities and separating them from the workday functions with which they conflict. In this way a lively new "zone of amusements" could be created and instead of roadside slums these approaches could be transformed into "avenues of gaiety and brilliance."[33]

A prominent portion of the mixed public found cruising on highway strips and Main Streets of America are adolescents. How they perceive and use the strip for their own purposes has recently been studied by Theodore Goldberg.[34] He describes "cruising" as a newly developed ritual activity of adolescents, one that is found in towns and cities of all sizes.

> Although the activity consists of nothing more elaborate than driving down a popular boulevard watching other people on week-end nights, it is transformed into an animated social attraction of bumper-to-bumper traffic as hundreds of cars filled with youths flood into the same area.[35]

Here Goldberg is speaking of the strips observed in the San Francisco Bay Area, but the activity is the same in other places, though the numbers may vary. The teen-agers proceed down the strip, and when they reach the end of the activity area they turn around and drive back. Along the way they interact between cars and occasionally pull off at one of the parking hang-outs where "cruisers" stop to socialize. Anything from a hamburger stand to an isolated parking lot may serve as a gathering place. Goldberg's studies indicate that the "cruisers" on a strip are predominantly local teen-agers who use the strip as a social arena. This arena has developed as a result of social and cultural forces operating in the adolescent sub-culture—namely, the need to gather, the impact of the automobile, and the lack of adequate institutions to serve their group needs. Unfortunately, the teen-agers are only one of many groups using the strip, so conflicts with other members of the public arise. Goldberg proposes design solutions to alleviate such conflicts and provide for the needs of adolescents and the larger community.

The new highway-oriented landscape, with its proliferation of garish signs, parking lots, and drive-in establishments, represents a commercial response to the presence of a motoring public. It has also been characterized as a form of folk art reflecting popular tastes and desires. Venturi, Brown, and Izenour, for example, argue that we should be *"learning from Las Vegas,"* where such pop architecture is seen in its clearest form.[36] They emphasize the importance of symbolic and representational elements drawn from past experience and emotional association. In the Las Vegas type of landscape:

> . . .it is the highway signs, through their sculptural forms or pictorial silhouettes, their particular positions in space, their inflected shapes and their graphic meanings which identify and unify the megatexture. They make verbal and symbolic connections through space, communicating a complexity of meanings through hundreds of associations in a few seconds from far away. Symbol dominates space. Architecture is not enough. Because the spatial relationships are made by symbols more than by forms, architecture in this landscape becomes symbol in space, rather than form in space. Architecture defines very little. The big sign and the little building is the rule of Route 66.[37]

The studies of Jackson, Goldberg, Venturi, and their associates on highway strips all provide insights into the functioning of these areas because they focus on how people perceive and use them. Each author has described some of the functions served by the highway strip and indicated that a better design could be achieved by alterations in keeping with the underlying order perceived. There is little doubt that a more pleasing design is possible and that a grouping of establishments could achieve that end. It is not clear whether the present forms represent the preferences of the public, as this was not studied directly. Even if the current strips do reflect public preferences, it does not follow that they represent the best way to serve the public's needs, as the public may not be aware of any other alternatives. One could imagine, for example, that the cruising activity of adolescents described by Goldberg could be enhanced by certain alterations of the strip design. But if the basic need is for a social arena, this could also be provided in other forms, thus eliminating the need for "cruising." The important thing is to understand the behavior in the natural setting before attempting to introduce new design.

The most comprehensive body of research on human behavior in its natural setting has been carried out in a tiny town labeled Midwest, Kansas, which had a population of 830 on January 1, 1964.[38] Instead of a general flavor of small towns or a discussion of one type of small-town area, this research provides a systematic inventory of all the behaviors and settings that are found in a particular town.

Ecological psychology

An outstanding exception to the general tendency of psychologists to rely on data gathered in laboratories or clinics is the work of ecological psychologists.[39] In 1947 the Midwest Psychological Field Station was established in Oskaloosa, Kansas, by Roger G. Barker and Herbert F. Wright as a facility of the Department of Psychology, University of Kansas. Barker and Wright founded the Midwest Field Station to make ecological studies of human behavior in a natural psychological habitat. They observed that

> the descriptive, natural history, ecological phase of science which is so strongly represented in the biological sciences, sociology, anthropology, earth sciences, and astronomy has had virtually no counterpart in psychology. This has left a serious gap in psychological knowledge, for in leaving out ecological methods, psychology has almost completely omitted a basic scientific procedure that is essential if some fundamental problems of human behavior are to be solved.[40]

Since the conditions in which behavior usually occurs cannot be created artificially in the laboratory, it becomes necessary to investigate the behavior in the field, where it occurs. But this is not easy. Much of the work of Barker and Wright in the small town of Midwest has been devoted to

developing new concepts and field techniques for measuring, recording, and analyzing the behavior they observed. The ecological units identified are defined in a rigorous fashion and the records of behavior published as a permanent data base.

An early record of this sort is provided in *One Boy's Day,* which recounts what one boy, Raymond, seven years and four months old, did during one complete day, April 26, 1949.[41] What he did and what his home, school, neighborhood, and town did to him from the time he woke up until he went to sleep that night was recorded by a team of eight observers (all familiar to the boy). They included all observable behavior, vocalizations, and body movements as well as on-the-spot impressions and inferences of Raymond's perceptions, motives, and feelings. These data could then be analyzed in terms of behavior episodes and behavior settings.

The behavior episode of an individual was seen as the smallest unit of value for ecological studies. However, larger, extraindividual units were soon discovered. In trying to sample the total universe of behavior in Midwest, Barker and Wright used the usual stratification guides—age, sex, social class, race, education, and occupation. But they soon found that another way to find distinctive types of behavior was to take samples in such divergent places as the drugstore, Sunday school classes, 4-H meetings, and football games. In such places groups of people tend to provide consistent configurations of behavior that persist year after year despite constant changes in the persons involved. The discovery of these *"standing behavior patterns"* in Midwest was a crucial step in the identification of a suitable community unit for ecological studies.[42]

It became clear that most standing behavior patterns are attached to particular places, things, and times—that is, to parts of the nonbehavioral context of the town. In addition, there appears to be a perceived fittingness or "synomorphism" between the patterns of behavior and the attributes of the nonpsychological context to which they are anchored. Thus a baseball game fits a baseball diamond, eating is appropriate in cafés but not libraries, and walking and riding on streets and sidewalks conform to the directions of these routes. Abrupt changes in an individual's behavior occur as the subject leaves one such behavior setting and enters another. A *behavior setting* is defined as: "a standing behavior pattern together with the context of this behavior, including the part of the milieu to which the behavior is attached and with which it has a synomorphic relationship."[43] Some examples are Kane's Grocery, Howell's Hayloft, 4-H Club Picnic, Capitol City Paper Route, Presbyterian Church, Hooker Tavern and Pool Hall, Home Meals, Boy Scout Pop Stand, and Thanksgiving Day. This unit, the behavior setting, is the one used in studying the community of Midwest. Very little behavior in this town occurs outside the limits of such settings, for they blanket the town.

The results of a full-scale behavior setting survey of Midwest are reported in *Midwest and Its Children.* This survey may be thought of as a

psychological map of Midwest, on which are represented the areas of the town that have general behavioral significance to the citizens. The precisely defined, systematic methods of Barker and Wright enable them to present clear, quantitative measures of many aspects of people-environment interaction in Midwest. For example, they tell us that during the survey year 1951-1952 there were 2,030 behavior settings in Midwest, of which 1,445 were located within homes of the town, family behavior settings, and 585 were found in its more public behavior areas or community behavior settings.[44] Furthermore, they can tell us what proportion of the residents' time was spent in each type of behavior setting—i.e., one third of their waking time is spent in community settings and the remainder in family settings. This has also been tabulated according to age and in a later study compared with a town in Yorkshire, England.[45] Thus, for example, one can see in Table 4.3, taken from this study, that in both towns there is a gradual increase in territorial range from infancy to adulthood with a decrease in elderly people to about that of adolescents. However, differences can also be noted that indicate that American children have much greater access to their town's total stock of behavioral settings than their English counterparts. With further similar studies in diverse environments, the method could be useful in comparing rural-urban differences or even national character.

In addition to the range of settings and amount of time spent in each, it is possible to measure the degree of involvement and responsibility of each person's participation in a behavior setting. This is referred to as the *degree of penetration*. It can range from no active involvement in Zone 1, to Zone 6, the position of a single leader with authority over the whole setting and without whom it would not function (Figure 4.7). By using the penetration index, the maximum depth of penetration into each of Midwest's behavior settings was rated for each Midwest resident and each population subgroup. This measure clearly indicated the degree to which children in

Table 4.3 Population (P), Territorial Range (TR), and Territorial Index (TI) of Age Groups in Midwest and Yoredale

Age group	**Midwest**			**Yoredale**		
	P	**TR**	**TI**	**P**	**TR**	**TI**
Aged (65 years and over)	162	462	80	178	332	67
Adult (18-64 years)	375	578	99	770	491	99
Adolescent (12-17 years)	50	464	80	107	329	67
Older School (9-11 years)	26	389	67	51	274	55
Younger School (6-8 years)	28	359	62	72	251	51
Preschool (2-5 years)	50	363	63	81	214	43
Infant (under 2 years)	24	329	57	41	125	25
All ages	715	579	100	1300	494	100

SOURCE: Roger G. Barker and Louise S. Barker, "The Psychological Ecology of Old People in Midwest, Kansas, and Yoredale, Yorkshire," *Journal of Gerontology*, 16 (1961), 146. Reprinted by permission.

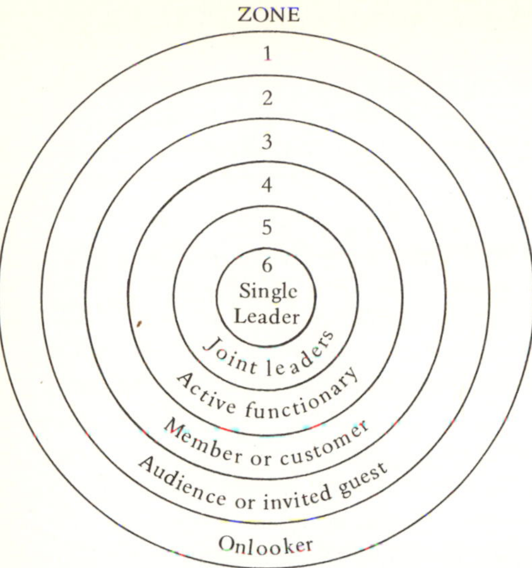

Fig. 4.7 Zones of penetration into behavior settings. (Roger G. Barker and Herbert F. Wright, *Midwest and its Children*, The Shoe String Press, Inc., Hamden, Connecticut, 1955, p. 70. Adapted by permission.)

Midwest were essential not only as onlookers but also as performers and members of settings.

By 1963, the Midwest Field Station had gathered eighteen day-long specimen records, such as the record of Raymond mentioned previously. These records served as primary data for a majority of investigations found in a later volume, edited by Barker, that focused on "*the stream of behavior*.[46] By carefully examining the total behavior stream, it is possible to identify larger environmental units. Schoggen identified such a unit, which he called an *environmental force unit* (EFU).[47] This is defined as "an action by an environmental agent toward a recognizable end state for a person."[48] In the eighteen records of the children analyzed by Schoggen, the median daily number of environmental force units was 386.5, which means that, during the course of a day, those around the median child engaged in that number of actions to direct, control, modify, support, or otherwise cope with the child's behavior.[49] This amounts to three environmental force units every five minutes. The numbers used here provide some impression of the precision possible with the methods employed, but, as Schoggen expresses it:

> Perhaps the one most significant result of this study lies in the demonstration that the social environment of the child, as recorded in specimen records, displays readily recognizable properties of directedness with regard to the child, i.e., the

social environment appears to have intentions with respect to the child which can be used for descriptive and analytical studies of environmental forces acting upon children in ordinary, everyday life.[50]

Some of the implications of the way environmental forces operate in behavior settings are developed in detail by Barker in *Ecological Psychology*, which marks the summation of twenty years of research at the Midwest Field Station.[51] This book and *Recording and Analyzing Child Behavior* by Wright encompass most of the field of ecological psychology as conceived by the authors at the time the books were written.[52] Barker explains in detail the theory of behavior settings, and their manner of operation is not unlike that of people-machine systems, considered earlier in Chapter 2. Barker explains:

> According to behavior setting theory, the ecological environment of human moral behavior and its inhabitants are not independent; rather, the environment is a set of homeostatically governed components, human components, and control circuits that modify the components in predictable ways to maintain the environmental entities in their characteristic states.[53]

The human components help to maintain the behavior in keeping with the particular behavior setting. Thus deviations may be kept in check, disruptive forces removed, and greater demands made on participants when necessary. This is most clearly illustrated in terms of undermanned and over or optimally manned behavior settings. When a setting has fewer than the optimal number of inhabitants to maintain its standing behavior pattern, it is an *undermanned setting*. In such behavior settings there is more force exerted on each inhabitant, as the standing behavior pattern must be accomplished by fewer people. The way it works in a school setting has been investigated in a volume by Barker and Paul V. Gump, *Big School, Small School*.[54] It was demonstrated that in small schools students report many more pressures to participate in various behavior settings as well as to perform in more responsible roles. In the undermanned setting, the average person has a greater degree of satisfaction. What is so fascinating here is the convincing manner in which the operation of these control mechanisms is worked out.

Recently Robert B. Bechtel and his associates applied the behavior setting survey technique in an urban area.[55] Two city blocks containing forty-six families were compared with Midwest. Bechtel stated that there was no major problem involved in transferring the technique to the urban environment. With some adjustments in methodology, the behavior setting survey can be used to provide precise, quantitative expressions of the differences between a small town and an urban neighborhood. For example, the behavior settings available to Bechtel's urban residents were nearly three times more numerous than those available to town residents. But the people in Midwest penetrated more deeply into their settings—that is,

they had more control over them than the city block inhabitants, who served mostly as onlookers or audience. The city environment is thus described by Bechtel as an overmanned condition likely to foster passivity and apathy. He suggests that many more undermanned settings would help. This would involve "decentralization, fragmentation of many efforts, and a new value system that replaces mechanical efficiency of the social structure as the highest goal with the necessity of participation of members as the highest good."[56] Newman's defensible space design clearly incorporates some of the features of undermanned or optimally manned settings, while huge, impersonal, high-rise projects may be regarded as overmanned. The apathy and indifference of the inhabitants of these overmanned settings make them more vulnerable to crime.

Unfortunately, the people who live in a neighborhood are generally not the ones who make the important decisions about their environment. In fact, there is often a major communication gap between the planners and the people involved. This is illustrated by reference to the West End of Boston, where a series of studies showed up some major failings of a federal urban renewal policy.

Cultural relativity of neighborhood perception

The distinctive differences in the outlooks of slum dwellers, or the lower class, or poor people and the lack of appreciation of these differences by urban renewal authorities or social reformers are a major reason for the failure of urban renewal programs.[57] According to Amos Rapoport, "this problem of subjective and cultural relativity of perception and cognition has come as something of a shock to the designers since they had never considered it."[58] Redevelopment and relocation procedures in large American cities frequently benefit the redeveloper and his tenants more than the site residents.[59] There is a communication gap between the urban renewal officials and the site residents. The former may see an area as a slum while the latter regard it as a pleasant neighborhood.[60] Even in cases where individual housing may be improved, the gain is often counterbalanced by the loss of community. These problems have all been thoroughly documented in relation to the urban renewal project in Boston's West End.

In this predominantly Italian area of lower- and working-class tenement residences, an ambitious five-year longitudinal study was carried out centering on the psychosocial characteristics of area residents both before and after their relocation.[61] Some of the earlier findings were reported in a book by Herbert Gans, who lived there as a participant-observer and conducted formal and informal interviews with many West Enders as well as redevelopment officials.[62] The title of his book, *The Urban Villagers,* indicates the parochial nature of the life of the inhabitants of this high-density urban neighborhood. The distinctive nature of their subculture

and the way in which its values helped maintain the identity of its members have been outlined here and elsewhere.[63] Such subcultures survive and thrive in spite of the "melting pot" myth. Disapproval will not make them disappear. To plan for them properly, understanding is necessary.

The depth and thoroughness of the West End study are exemplified in a follow-up study of the former West Enders some two years after relocation that is described by Marc Fried in a much-quoted article, "Grieving For a Lost Home."[64] The loss of home and neighborhood had such intense reactions among the people forced out of the West End of Boston that Fried felt they could most accurately be described as grief. Among the commonly occurring symptoms were

> . . . feelings of painful loss, the continued longing, the general depressive tone, frequent symptoms of psychological or social or somatic distress, the active work required in adapting to the altered situation, the sense of helplessness, the occasional expressions of both direct and displaced anger and tendencies to idealize the lost place.[65]

Remarkable is the tenaciousness of the imagery and the effect, for the follow-up interviews took place two years after relocation, a period sufficiently long for most people to have made a new adjustment. One of the most evocative questions that seemed to stir up sad memories even among those who had denied feeling sad or depressed was, "How did you feel when you saw or heard that the building you had lived in was torn down?"

Fried explained the nature of the loss brought about by relocation in terms of two major sets of factors, the spatial identity and the group identity. The group identity depended more on the feelings toward friends and neighbors who lived in the area. But separate from this was the spatial identity, which will be discussed in more detail since it is more often overlooked.

As in any other grief reaction, the spatial factor in the grief reaction of relocated West Enders results from a disruption in their relationship to the past, present, and future. Such losses change one's routines, relationships, and expectations. This can result from an alteration of the world of physically available objects and spatially oriented objects. Not only is the residential area one in which a vast and interlocking set of social networks is localized, but the physical area itself can be considered an extension of home:

> Even familiar and expectable streets and houses, faces at the window and people walking by, personal greetings and impersonal sounds may serve to designate the concrete foci of a sense of belonging somewhere and may provide special kinds of interpersonal and social meaning to a region one defines as "home."[66]

Fried goes as far as to say that a sense of spatial identity is fundamental to human functioning and, at least in the working class, is tied to a specific

place. The intensity of grief felt by West Enders relocated elsewhere was a function of their feelings for the area. The greater the person's prelocation commitment to the area, the more likely was a reaction of severe grief. Similarly, the greater the area of the West End known by the individual, the more likely was a strong response of grief. The most parochial of all, those who knew only their own block did not have as great an attachment to the area as those familiar with larger portions of the West End. Furthermore, although familiarity with the region is related closely to extensiveness of personal contact, "the spatial patterns have independent significance and represent an additional basis for a feeling of commitment to that larger, local region which is 'home.' "[67]

Fried emphasizes that this pattern of strong attachment to place is particularly strong among the working class. However, it might be fruitful to investigate the universal characteristics involved. It seems likely that people of all types and classes develop a strong attachment to place. Within the city of Boston, Firey noted earlier the fierce attachment of upper-class residents to Beacon Hill, revealed by their continuous, often expensive struggle against business encroachments and apartment-hotel developments.[68] It seems worthwhile to try to investigate this type of attachment more carefully in all sorts of settings, to determine what needs must be provided for in the largely artificial environment of the years ahead.

Alvin Toffler, in *Future Shock*, describes the increasing tendency toward more transient relationships with places.[69] Yet we know little of the consequences of such a trend. Melvin M. Webber and others have argued that changes in technology may provide opportunities for communication to people widely scattered in the urban landscape. Instead of depending on neighbors, people may now use automobiles to make contact with friends or relatives elsewhere in the city, so that "community without propinquity" is possible.[70] Michelson agrees with Webber insofar as he speaks of the cosmopolitan group who "claim the whole city as their empire" but points out that there are also people who are primarily citizens of their local area, with limited horizons.[71] He has found great differences between these two types of people in their perception of the ideal urban environment. What proportion of the people belong in each group now and in the future may have important planning consequences for city form.

A community-attitudes survey prepared for the Royal Commission on Local Government in England provides some evidence of the proportion of people there who acknowledge an attachment to place and some idea of the size of this home area.[72] It was based on fully representative samples of 2,199 electors chosen from 100 local authority areas throughout England, excluding London. Two of the main aims of the survey were to determine whether people feel there was a community area to which they belonged, and, if so, what was the physical size of that area. The ultimate aim was to compare such information with the boundaries and divisions of local government authorities. The first question posed was, "is there an area

around here, where you are now living, that you would say you belong to, and where you feel at home?" Nearly four fifths of those interviewed did claim some feeling of attachment to a home community area, and the likelihood of such a feeling increased with length of residence in the area. To determine the physical extent of the perceived home area, which, incidentally, was much the same for those who felt some attachment as for those who didn't, each respondent's verbal description of the area was utilized by the place and street names volunteered. A major finding was that the community area thus defined is quite small. In most cases it is considerably smaller than the local government areas in which the people reside. In urban areas the majority define its extent as approximately the size of a group of streets or smaller. Further evidence on size of perceived neighborhoods is provided by Terence Lee, who has identified a hierarchy of neighborhoods.[73]

The urban neighborhood as a sociospatial schema

Lee believes that the "the duality of physical and social neighborhoods can be joined only by a phenomenological approach."[74] He had a representative sample of homemakers in a small, provincial English city outline on a map the extent of their neighborhood and describe in detail their behavior in the immediate environment. In his analysis of the results, he states that perhaps the most accurate way to consider the residents' perceptions of their urban neighborhood is as a "sociospatial schema." A threefold typology of neighborhood schemata was generalized from the data collected. Lee labels these as the social acquaintance neighborhood, the homogeneous neighborhood, and the unit neighborhood.

The *social acquaintance neighborhood* is a small physical area whose boundaries are set by social interaction. Within it everybody knows everybody else, whether they have anything to do with each other or not. The *homogeneous neighborhood* boundaries are set by the size, price, or condition of the houses and the kinds of people who live in them. Some friendships in nearby streets develop over time from a rather large number of acquaintances. However, the most pervasive social relationship is one of "mutual awareness," which Lee considers likely to exert social control despite a lack of overt interaction. The *unit neighborhood* is generally larger than the others and contains a balanced range of amenities, such as, shops, schools, churches, and clubs. It corresponds most closely to the planner's conception of the neighborhood unit, which has recently been subject to some disenchantment.[75]

To test the effect of differing physical environments upon the neighborhood, Lee compared the schemata with the contents of the larger locality, defined as a half-mile radius from the subject's house. His results led him to believe that the locality remains a vital unit for city planning. He found

Fig. 4.8 Individual neighborhood schema. (Terence R. Lee, "Do We Need a Theory," in *Architectural Psychology*, ed. David V. Canter, RIBA Publications, London, 1970, p. 22. Adapted by permission.)

that it is an area where regular shopping trips occur and where the majority of individuals have friends and club memberships. A Neighborhood Quotient (Nh. Q.) was worked out to express the individual schema as a ratio of the locality. Figure 4.8 shows examples of individual schemas. These were variable, both in terms of size and in degree of participation. In spite of great diversity, there are also some uniformity and overlapping of individual neighborhoods. However, it is probable that there are no definite limits acknowledged by all. Instead, gradients and focuses of interaction may appear, perhaps with concordance on prominent boundaries that might be greater if they were accentuated. The evidence leads Lee to conclude that planning should be directed toward heterogeneous physical and social layouts deliberately emphasizing the local satisfaction of needs. In the field of specific design Lee suggests that open spaces and traffic roads be used to delineate clearly subunits and that shops be positioned off center toward the downtown edge of the unit.

The threefold typology of neighborhoods developed by Lee in England confirms previous findings of Gaston Bardet based on a study of some sixty rural and urban places in Europe and Africa.[76] Bardet distinguished three local groupings that he called the "patriarchal, domestic, and parish 'echelons' " or degrees. Georgia Zannaras in Ohio[77] and Alain Metton in Paris[78] have gathered similar data on neighborhoods. These data provide informa-

tion that may help planners avoid the kind of situation described by Daniel F. Doeppers that occurred in the Globeville neighborhood in Denver.[79] Here a freeway was constructed that cut through the social fabric of an old ethnic neighborhood and contributed to its decline.

Neighborhoods as territories

A study by Frederick Boal in Belfast, Northern Ireland, shows how abruptly the borderline between neighborhoods can be defined by the activity patterns of residents of this troubled city.[80] He interviewed people in the Shankill-Falls area of the city, where the Protestant and Roman Catholic groups are residentially segregated from each other but in close spatial proximity. Figures 4.9 and 4.10 show the location of the study area in relation to the pattern of street decorations for the Orange (Protestant) July Twelfth celebrations, and households where at least one member has a surname or first name that could be classified as "Roman Catholic." The two patterns are almost reversed images of each other. The study area contains two distinct territories separated by the "divide" of Cupar Street. The people on opposite sides of this divide live in entirely different mental worlds. They are of different religions and read different newspapers. Furthermore, their activity patterns are almost totally separated, with no socially meaningful connections linking the two portions of the study area in spite of their proximity. The strict observance of territorial limits in

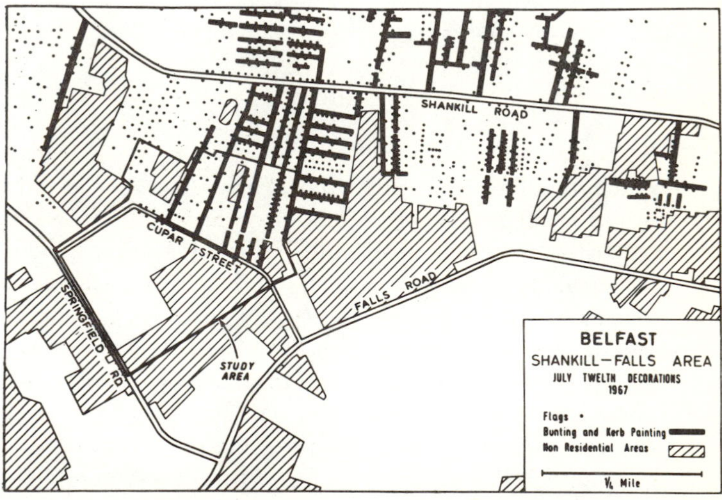

Fig. 4.9 Belfast: July 12th Decorations, Shankill-Falls Area, 1967. (F. W. Boal, "Territoriality on the Shankill-Falls Divide, Belfast," *Irish Geographer,* 1969, VI, 34-35. Reprinted by permission.)

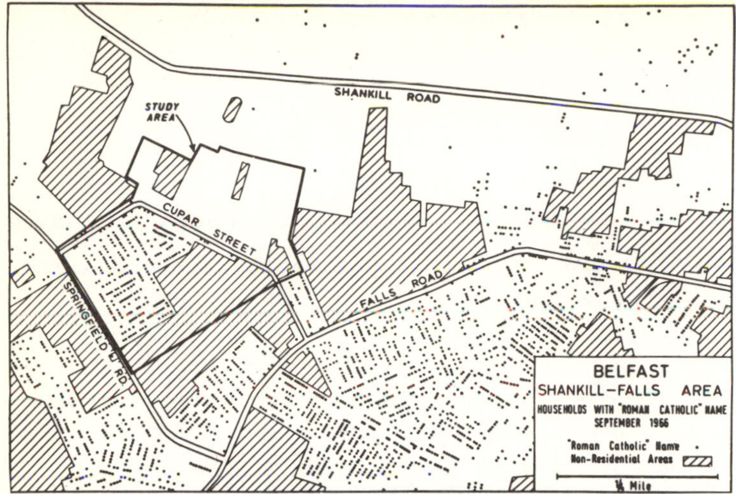

Fig. 4.10 Belfast households with Roman Catholic names, Shankill-Falls Area, 1966. (F. W. Boal, "Territoriality on the Shankill-Falls Divide, Belfast," *Irish Geography*, 1969, VI, 34-35. Reprinted by permission.)

Belfast no doubt serves to reduce the amount of direct confrontation and conflict. The study provides a striking example of the importance of the largely invisible landscape that may exist in people's minds and condition their behavior and activity patterns, yet not be readily discernible in physical form to the outside observer. This raises questions about the maps people carry in their heads and how these are developed—issues that are considered in more detail in the following chapter.

Summary

In the preceding discussion we noted many links between the topics at the level of small town and neighborhood and other scales. We saw some similarities between site planning and the social behavior and seating arrangements in face-to-face groups, as both emphasize to some degree the effect on the communication process of direct visual contact. The work of Jackson and of the psychological ecologists indicates how diverse the settings within a small town can be and how full of meaning for those involved in each behavioral setting. We compared behavioral settings with people-machine systems in terms of the way the individual's behavior served to stabilize the total system, which consists of human components, nonhuman components, and control circuits. The research of the ecological psychologists demonstrated the interactions of the various components in a people-environment system. But the value of viewing people-

environment interactions as a system has been emphasized at each of the scales considered in previous chapters. Similarly, the value of naturalistic observation of real-life settings, as seen, for example, in the work of Jackson and Goldberg, was previously noted in research on room geography and architectural space. In this chapter we discussed in detail the extremely thorough ecological approach of Barker and his associates to emphasize the type of concepts that must also be developed and the power of such concepts to explain what is observed. Examples are the concepts of undermanned and overmanned behavior settings. These could serve admirably to explain the differences in behavior in the various types of housing projects investigated by Newman and to illustrate why "defensible space" design might work.

An important issue raised was the communication gap between planners and the people most affected by a plan. Not all planning leads to improvements. Some projects may create worse problems than those they set out to solve. This is especially true when stereotypic solutions are imposed from the outside. The real test of a plan comes when it is measured by the yardstick of the users. Unfortunately, slum dwellers are not always consulted before slum clearance decisions are made. Similarly, in other planning situations the views of those most affected may never be canvassed. This was noted in the chapter on personal space, and it will be brought up again in Chapter 6, where the question of the attitudes of the major decision-makers is raised. Even when an attempt is made to bridge the gap between the planners and the people, the process is not simple, for often there are great cultural differences that make easy adjustments unlikely. This is well illustrated by some of the conclusions of Donald Appleyard based on his experience with the Venezuelan urban and regional development project centering on Ciudad Guayana. Speaking of the designer, he states:

> his motivations are general, diffuse, and future oriented, whereas the inhabitant's are usually particular, specific, and present oriented; his experience with cities is usually much greater than that of the inhabitant, which makes it difficult for him to see a city or plan with an "innocent" eye. His familiarity with the city is usually more extensive than intensive, and his information media are so abstracted and amplified with "objective" data that his world tends to become divorced from the real city. His very abilities and knowledge create the gap.[81]

In urban neighborhoods the situation is further complicated by the addition of other decision-makers, such as absentee property owners. The communication gaps separating the various parties concerned can lead to conflicts in locational decisions.[82]

Although there has been some doubt expressed about the viability of the conventional neighborhood concept, evidence from perception studies indicates that a majority of those interviewed have definite notions of their

neighborhood. There are great individual differences in the perceived size of the neighborhood and the degree of attachment to it, but some general agreement may appear in the ways the phenomenal neighborhoods overlap. It might be useful to consider neighborhoods in the context of territoriality and home range. Thus, as the studies of the psychological ecologists have indicated, one could see the very young and the very old as more restricted in range and more closely tied to territory. The same might be said of the poor. Territoriality may also be highly developed in situations of intergroup conflict, such as that in Belfast described by Boal. There, a very clear boundary line separates the activities of the contending parties, though an outsider would notice nothing unusual in the landscape.

In this chapter *environment* was mainly the built environment and the social environment. The discussion of physical determinism in the sections on site planning and social behavior and defensible space exemplifies a strong focus on the built environment. On the other hand, the social or psychological environment of a small town was inventoried in the section on ecological psychology, where we saw how the physical setting and human actions combine to create behavior settings. In addition, another, largely symbolic, environment appeared with greater prominence than in preceding chapters. Thus, for example, Jackson and Venturi and his associates emphasized the symbolic aspects of the new pop architecture of highway strips. Boal's article indicated the presence of an invisible landscape in people's minds that created territorial divisions. Territoriality also appeared at the scales of the preceding chapters. Certain other aspects of symbolism also appeared in earlier chapters where environmental features were charged with social or psychological meanings. But concern with symbolism and mental maps becomes increasingly important as larger, more abstract areas are discussed. This will become evident as we move to the scales of the city, the region, the country, and the world in succeeding chapters.

Notes

1. See, for example, the discussion of basic human needs and the kinds of environments that are assumed to satisfy them in Benton MacKaye, *The New Exploration,* University of Illinois Press, Urbana, Ill., 1962. First published by Harcourt Brace Jovanovich, Inc., New York, 1928.
2. Robert Gutman, "Site Planning and Social Behavior," in R. W. Kates and J. F. Wohlwill, eds., "Man's Response to the Physical Environment," *The Journal of Social Issues,* 22, No. 4 (October, 1966), 103-115; also reprinted in Proshansky *et al., Environmental Psychology,* pp. 509-517.
3. For a research report and review of the effects of housing on health and performance, see Daniel M. Wilner and Rosabelle Price Walkley, "Effects of Housing on Health and Performance," Chapter 16 in Leonard J. Duhl, ed., *The Urban Condition,* Simon & Schuster, Inc., New York, 1963, pp. 215-228.
4. William H. Whyte, Jr., *The Organization Man,* Jonathan Cape, London, 1957, p. 330.

5. *Ibid.*, p. 330.
6. *Ibid.*, p. 332.
7. See the discussion of all such factors in William Michelson, *Man and His Urban Environment: A Sociological Approach,* Addison-Wesley, Publishing Co., Inc., Reading, Mass., 1970, Chapter 8.
8. *Ibid.*
9. Robert Athanasiou and Gary A. Yoshioka, "The Spatial Character of Friendship Formation," *Environment and Behavior,* 5, No. 1 (March, 1973), 43-65.
10. *Ibid.*, p. 61.
11. *Ibid.*, p. 60.
12. *Ibid.*, p. 62.
13. Oscar Newman, *Defensible Space,* MacMillan Publishing Co., Inc., New York, 1972.
14. *Ibid.*, p. 1.
15. *Ibid.*, p. 3.
16. *Ibid.*, p. 10.
17. Jane Jacobs, *The Death and Life of Great American Cities,* Random House, Inc., New York, 1961, Chapter 2.
18. For a collection of articles written by J. B. Jackson, see Ervin H. Zube, ed., *Landscapes: Selected Writings of J. B. Jackson,* University of Massachusetts Press, Amherst, 1970.
19. J. B. Jackson, "The Almost Perfect Town," *Landscape,* 2, No. 1 (Autumn, 1952), in *Landscapes,* ed. Zube, pp. 116-131; quotation, p. 116.
20. *Ibid.*, p. 125.
21. *Ibid.*, p. 122.
22. *Ibid.*, p. 122.
23. *Ibid.*, p. 123.
24. *Ibid.*, p. 128.
25. *Ibid.*, p. 128.
26. Pierce F. Lewis, "Small Town in Pennsylvania," *Annals of the Association of American Geographers,* 62, No. 2 (June, 1972), 323-351.
27. *Ibid.*, p. 327.
28. *Ibid.*, p. 327.
29. J. B. Jackson, "Other-directed Houses," *Landscape,* 6, No. 2 (Winter, 1956-1957); also in Zube, ed., *Landscapes: Selected Writings of J. B. Jackson,* p. 55-72.
30. J. B. Jackson, "The Social Landscape," in Zube, ed., p. 149.
31. J. B. Jackson, "Other-directed Houses," in Zube, p. 62.
32. J. B. Jackson, "The Social Landscape," in Zube; p. 149.
33. J. B. Jackson, "Other-directed Houses," p. 72.
34. Theodore Goldberg, "The Automobile: A Social Institution For Adolescents," *Environment and Behavior,* 1, No. 2 (December, 1969), 157-185.
35. *Ibid.*, p. 164.
36. Robert Venturi, Denise Scott Brown, and Steven Izenour, *Learning From Las Vegas,* The M.I.T. Press, Cambridge, Mass., 1972.
37. Venturi, Brown, and Izenour, "Learning From Las Vegas," in William H. Ittelson, ed., *Environment and Cognition,* Seminar Press, Inc., New York, 1973, p. 103.
38. Roger G. Barker, *Ecological Psychology: Concepts and Methods For Studying the Environment of Human Behavior,* Stanford University Press, Stanford, Calif., 1968, p. 92.
39. Ecological psychology could be considered a specialized development within the broader area of field theory in social psychology originated by the work of Kurt Lewin and his associates. See Morton Deutsch, "Field Theory in Social Psychology," Chapter 5 in Gardner Lindzey, ed., *Handbook of Social Psychology,* Addison-Wesley Publishing Co., Inc. Reading, Mass., 1954, pp. 181-222.

40. Roger G. Barker and Herbert F. Wright, *Midwest and Its Children,* Row, Peterson and Company, Evanstan, Ill., 1955, p. 1.
41. Roger G. Barker and Herbert F. Wright, *One Boy's Day,* Harper & Row Publishers, New York, 1951.
42. Barker and Wright, *Midwest and Its Children,* p. 8.
43. *Ibid.,* p. 9.
44. *Ibid.,* p. 84.
45. Roger G. Barker and Louise S. Barker, "The Psychological Ecology of Old People in Midwest, Kansas, and Yoredale, Yorkshire," *Journal of Gerontology,* 16, No. 2 (April, 1961), 144-149.
46. Roger G. Barker, *The Stream of Behavior: Explorations of Its Structure and Content,* Appleton-Century-Crofts, New York, 1963; see Appendix 1.1 for sources of specimen records.
47. Phil Schoggen, "Environmental Forces in the Everyday Lives of Children," Chapter 3 in *The Stream of Behavior,* ed. Barker, pp. 42-69.
48. See Barker, *Ecological Psychology,* p. 152, for definition.
49. Schoggen, "Environmental Forces in the Everyday Lives of Children," p. 49.
50. *Ibid.,* p. 69.
51. See Footnote 38.
52. Herbert F. Wright, *Recording and Analyzing Child Behavior,* Harper & Row Publishers, New York, 1967.
53. Barker, *Ecological Psychology,* p. 186.
54. Roger G. Barker and Paul V. Gump, *Big School, Small School,* Stanford University Press, Stanford, Calif., 1964.
55. Robert B. Bechtel, "A Behavioral Comparison of Urban and Small Town Environments," in Archea and Eastman, eds., *EDRA Two,* pp. 347-353.
56. *Ibid.,* p. 352.
57. For a good discussion of many aspects of this problem, see Leonard J. Duhl, ed., *The Urban Condition: People and Policy in the Metropolis,* Simon & Schuster, Inc., New York, 1963.
58. Amos Rapoport, "Observations Regarding Man-Environment Studies," *Man-Environment Systems* (January, 1970), 2.
59. For a discussion of this point, see Herbert J. Gans, "The Human Implications of Current Redevelopment and Relocation Planning," *Journal of the American Institute of Planners,* 25, No.1 (February, 1959), 15-25; reprinted in Herbert J. Gans, *People and Plans,* Basic Books, New York, 1968.
60. Marc Fried and Peggy Gleicher, "Some Sources of Residential Satisfaction in an Urban Slum," *Journal of the American Institute of Planners,* 27, No. 4 (November, 1961), 305-315; also in Proshansky *et al.,* pp. 333-346. See also Chester W. Hartmen, "Social Values and Housing Orientations," *Journal of Social Issues,* 19 (1963), pp. 113-131.
61. See Duhl, ed., *The Urban Condition,* p. 95.
62. Herbert J. Gans, *The Urban Villagers,* Glencoe Press, New York, 1962.
63. See the chapters in Duhl, ed., *op. cit.,* by Peter Marris, "A Report on Urban Renewal in the United States," pp. 113-134; Edward J. Ryan, "Personal Identity in an Urban Slum," pp. 135-150; and A. B. Hollingshead and L. H. Rogler, "Attitudes Towards Slums and Public Housing in Puerto Rico," pp. 229-245. A review of the sociological literature on this topic is found in William Michelson, *Man and His Urban Environment: A Sociological Approach,* Addison-Wesley Publishing Company, Reading, Mass., 1970, Chapter 3, "Life Style and Urban Environment." The anthropological approach may be sampled in Charles A. Valentine, *Culture and Poverty,* University of Chicago Press, Chicago 1968; Gerald D. Suttles, *The Social Order of the Slum: Ethnicity and Territory in the Inner*

City, University of Chicago Press, Chicago, 1968; and Ulf Hannerz, *Soulside: Inquiries into Ghetto Culture and Community*, Columbia University Press, New York, 1969.

64. Marc Fried, "Grieving For a Lost Home," in Duhl, ed., *The Urban Condition*, pp. 151-171.
65. *Ibid.*, p. 151.
66. *Ibid.*, p. 154.
67. *Ibid.*, p. 156.
68. Walter Irving Firey, "Sentiment and Symbolism as Ecological Variables," *American Sociological Review*, 10, No. 2 (April, 1945), 140-148.
69. Alvin Toffler, *Future Shock*, Random House, Inc., New York, 1970; see Chapter 5, "Places: The New Nomads."
70. Melvin M. Webber, "Order in Diversity: Community without Propinquity," in Lowdon Wingo, Jr., *Cities and Space*, Johns Hopkins Press, Baltimore, 1963, pp. 23-54; also in Proshansky *et al.*, *Environmental Psychology*, pp. 533-549.
71. William Michelson, *Man and His Urban Environment*, p. 90.
72. Royal Commission on Local Government in England, *Research Studies 9 Community Attitudes Survey: England*, Her Majesty's Stationery Office, London, 1969.
73. Terence Lee, "Urban Neighborhood as a Socio-Spatial Schema," *Human Relations*, 21, No. 3 (1968), 241-267; also in Proshansky *et al.*, *op. cit.*, pp. 349-370.
74. *Ibid.*, p. 351.
75. See, for example, Suzanne Keller, *The Urban Neighborhood: A Sociological Perspective*, Random House, Inc., New York, 1968.
76. Gaston Bardet, "Social Topography: An Analytico-Synthetic Understanding of the Urban Texture," in George A. Theodorson, ed., *Studies in Human Ecology*, Row, Peterson and Company, Evanston, Ill., 1961, pp. 370-383.
77. Georgia Zannaras,"An Empirical Analysis of Urban Neighborhood Perception," unpublished master's thesis, Ohio State University (geography), 1968.
78. Alain Metton, "Le Quartier: Étude Géographique et Psycho-Sociologique," *Canadian Geographer*, 13, No. 4 (Winter, 1969), pp. 199-316.
79. Daniel F. Doeppers, "The Globeville Neighborhood In Denver," *The Geographical Review*, 57, No. 4 (October, 1967), 506-522. For discussions of this type of problem, see also Gordon Fellman and Barbara Brandt, "Working-Class Protest Against An Urban Highway: Some Meanings, Limits, and Problems," *Environment and Behavior*, 3, No. 1 (March, 1971), 61-79, and, by the same authors, "A Neighborhood a Highway Would Destroy," *Environment and Behavior*, 2, No. 3 (December, 1970), 181-301.
80. F. W. Boal, "Territoriality on the Shankill-Falls Divide, Belfast," *Irish Geography*, 6, No. 1 (1969), 30-50.
81. Donald Appleyard, "City Designers and the Pluralistic City," Chapter 23 in Lloyd Rodwin, ed., *Planning Urban Growth and Regional Development: The Experience of the Guayana Program of Venezuela*, M.I.T. Press, Cambridge, Mass., 1969, p. 450. See also the chapters by William Porter, "Changing Perspectives on Residential Area Design"; Lisa Peattie, "Social Mobility and Economic Development"; Arthur L. Stinchcombe, "Social Attitudes and Planning in Guayana"; Donald Appleyard, "City Designers and the Pluralistic City"; and Lisa Peattie, "Conflicting Views of the Project: Caracas Versus the Site."
82. This problem is discussed in Julian Wolpert, Anthony Mumphrey, and John Seley, *Metropolitan Neighborhoods: Participation and Conflict Over Change*, Commission on College Geography, Resource Paper No. 16, Association of American Geographers, Washington, D.C., 1972.

Five

The city

"There seems to be a public image of any given city which is the overlap of many individual images."
Kevin Lynch

The city is perhaps the most complex and concentrated of our artifacts. The way we use personal space, rooms, buildings, groups of buildings, and neighborhoods can all be observed in the city. Yet the city is more than the sum of its parts, for it has a vitality and organization that transcends them all. Similar to an organism, to which it has often been compared,[1] it grows and passes through stages, and its parts are replaced even as the whole continues. One of the main obstacles preventing comprehension of cities today is the rapidity of this growth and change. Not only are there more and larger cities than ever before, but their internal structure is continually changing.[2] Changes in technology have led to shifts in the density and dominance of various portions.

All these shifts in the size, density, and internal arrangements require adjustments by the human inhabitants. But the behavioral implications are at best only dimly understood. Planners still do not know how to design an optimum environment with "man as the measure."[3] Furthermore, the situation is not likely to improve soon, for cities continue to evolve on a hit-or-miss basis with no clearly formulated goals. As C. A. Doxiadis has said, "By his actions man has created a great laboratory of human settlements in which he is both the research director and the guinea pig."[4] The

task of planners is to pay attention to what is happening to people in this laboratory to improve conditions. One might then ask which goals are to be used in assessing the improvement.

Cities can be built to serve any idea of "the good life" that can be proposed, but generally the goals are not stated. A stimulating exception to this tendency is found in *Communitas* by Paul and Percival Goodman,[5] which exposes the implicit assumptions of a number of modern plans and develops the planning implications of three radically different community models in a wise, witty, and provocative manner. The models selected for discussion were: the city closest to the current American pattern, the city of efficient consumption (a community aimed at eliminating the difference between production and consumption), and another that provided planned security with minimum regulation. A more recent effort to illustrate such planning from first principles is a paper by Christopher Alexander, "The City as a Mechanism for Sustaining Human Contact."[6] Its starting point is the hypothesis that "a society can be a healthy one only if each of its individual members has three or four intimate contacts at every stage of his existence."[7] From there, Alexander proceeds to show how to go about designing a city that creates the necessary conditions for such a state of affairs. More important than the question of whether his hypothesis or solution is correct is the underlying idea that city planning should start with clearly formulated goals for which design solutions are sought.

The type of research reported in this chapter may help to suggest appropriate goals and solutions by providing a deeper understanding of people's perceptions of and behavior in the urban environment. The final part of Chapter 4 was a discussion of how neighborhood boundaries may be defined by the activity patterns of the residents. Here we will resume the discussion of activity patterns, but we will start from the perspective of the neighborhood residents' views of the entire city, the larger environment within which the neighborhood is embedded.

Urban activity patterns

The geographers, Frank Horton and David Reynolds, studied the variations in action spaces of urban residents in Cedar Rapids, Iowa. *Action space* was defined as "the collection of all urban locations about which the individual has information and the subjective utility or preference he associates with these locations."[8] It was found that the action space of individuals is dependent on the location of the neighborhood and the location of their former residences. Familiarity with urban areas is linked to the individuals' activity patterns, which focus on their homes.

Horton and Reynolds conducted home interviews with residents of two Cedar Rapids residential areas—Cedar Hills, a recently developed middle-class suburban community located on the urban fringe, and Oak

Hill, an old, low-income central-city area. All the respondents were given a map of the Cedar Rapids metropolitan area and were asked to indicate on a five-point scale their familiarity with each of 27 delimited areas on the map. The data gathered were factor-analyzed to determine the basic dimensions of the group-action spaces, which are shown in Figures 5.1 and 5.2. In both cases there were systematic variations in degree of familiarity.

The suburban group, the Cedar Hills sample, was unfamiliar with several sections of the city. These, as shown in Figure 5.1, are: a large area of decreasing familiarity stretching south from the city (Factor 1), a northern area of low levels of familiarity (Factor 2), a relatively inaccessible area of low population density in the northwest (Factor 5), and an area of decreasing familiarity west of the central business district (CBD) (Factor 6). The map for the suburban group indicates two areas with which they were more familiar. The first (Factor 3) is an area along the major transportation axis of the city between the CBD and the major shopping plaza. The second is their home area. The suburban group's action space seems to be closely related to their activity patterns, such as traveling to work or shopping.

The central city residents, the Oak Hill sample, display a more concentric pattern of degrees of familiarity (Figure 5.2). They are most familiar with the home area (Factor 2) and moderately familiar with the area immediately

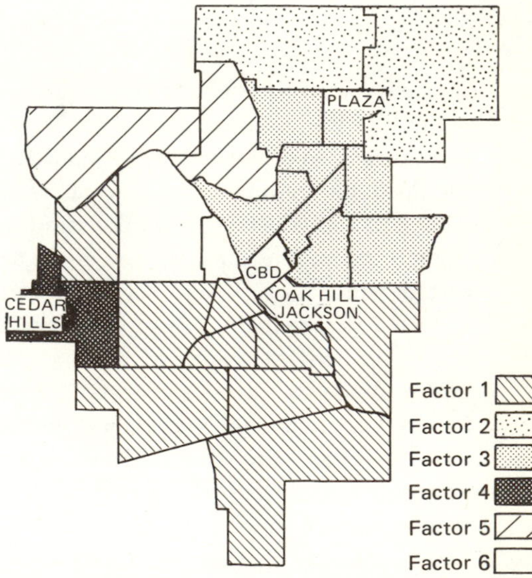

Fig. 5.1 Familiarity factor structure, Cedar Hills sample. (Frank E. Horton and David R. Reynolds, "Action Space Differentials in Cities," in *Perspectives in Geography: Models of Spatial Interaction*, eds. H. McConnell and D. W. Yaseen, Northern Illinois University Press, DeKalb, Illinois, 1971, p. 96. Adapted by permission.)

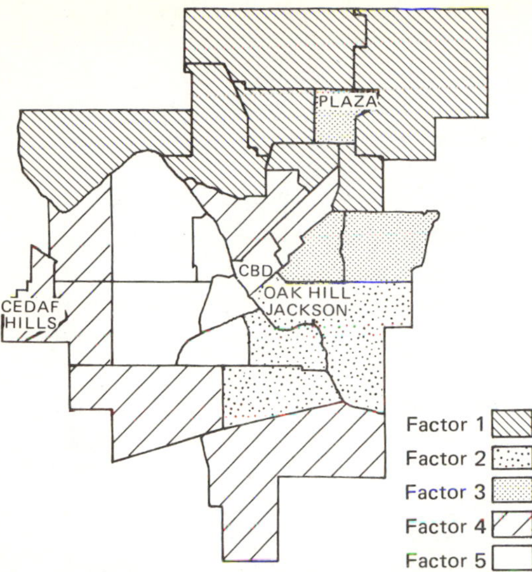

Fig. 5.2 Familiarity factor structure, Oak Hill sample. (Frank E. Horton and David R. Reynolds, "Action Space Differentials in Cities," in *Perspectives in Geography: Models of Spatial Interaction*, eds. H. McConnell and D. W. Yaseen, Northern Illinois University Press, DeKalb, Illinois, 1971, p. 96. Adapted by permission.)

adjacent on the north (Factor 3). For them there is also an area of decreasing familiarity west of the CBD (Factor 5). Surrounding these inner areas is a concentric ring of unfamiliar residential areas (Factor 4), beyond which is the northern area, the least familiar of all (Factor 1).

In both cases the map of familiarity of urban residents was clearly related to their location in the city. The home areas were well known, as well as certain other areas where the daily activity of the residents would be expected to take them. Beyond such areas was a larger portion of the city that was much less familiar to the residents. The less familiar areas were the ones outside the zones of the residents' regular activities. Horton and Reynolds state that, to understand the long-range spatial restructuring of the city, we must be aware of action space formation. It is not enough to know the objective spatial structure, as spatial decisions are not based on perfect knowledge.

It seems clear that insight into the planning of a city or its parts could be gained by direct investigation of urban activity systems. F. Stuart Chapin, Jr., and Richard K. Brail, for example, conceive of the metropolitan community as "a variety of entities—persons, firms, voluntary organizations, churches, governments, and so on—interacting in various ways in the pursuit of their everyday affairs."[9] Each of the entities may break down

into a series of subsystems with its own particular class of activities. J. Rannels, on one level, has discussed the kinds of linkages between various types of establishments in his book, *The Core of the City.*[10] On other levels, J. Douglas Porteous has presented a map of the activity system or territorial range of a teen-age gang in Victoria, British Columbia[11]; and the Horton and Reynolds study has analyzed action spaces of neighborhood residents.

The Rannels study examines activity patterns of establishments, defined as "individuals, or groups using a definite location as a recognizable place of business, residence, government or assembly."[12] Linkages between establishments are created by the interchange of people, goods, or information.

> The net balance of pulls exerted on *each* establishment by its linkages with others is a major factor in the spatial arrangement of land uses, so that each new establishment tends to locate where the forces of its expected linkages will be in equilibrium.[13]

A study of the linkages would provide measures of the activities engaged in by each establishment. Planners could then match the linkages of all types with the means for effectuating these linkages at various city locations, as in Table 5.1.

At the level of the social group, Chapin and Brail report the results of a large national survey of activity patterns of urban residents.[14] The analysis attempts to identify factors that account for various classes of weekday activity. It appears that significant differences are identifiable according to sex, socioeconomic class, and stage in life cycle. The authors propose to explore these differences further in relation to the way the various groups use their free time. Such information would enable the planner to apportion the community resources according to criteria other than the more commonly used economic demand or traffic counts.

Urban activity patterns have both temporal and spatial aspects. Thus daily, weekly, monthly, seasonal, annual, and even life cycles may be seen. Paul-Henry Chombart De Lauwe provides many revealing examples of how they operate in various portions of Paris.[15] The importance of certain neighborhood cinemas in the weekly cycle of young people was observed. Some individuals regularly attended the cinema closest to their home each Friday night for as long as two or three years. Thus the themes of the movies shown there serve as practically the sole source of thought and discussion for such groups. The morning bus to work may unite certain groups and create for them a regular social rhythm. In many areas one of the most significant spots is the local café, which, like urban neighborhood shops, as Jane Jacobs noted, may serve a variety of friendly functions.[16] The working-class restaurants near factories, sports arenas, beauty parlors, pools, gymnasiums, and various organizations may also serve as regular

Table 5.1 Classification Outline: Space Needs of Users and Space Provided by Buildings

Space as used by establishments	Usable space in buildings
I, II. *Kind of Business* (of the firm to which establishment belongs) such as: Manufacturing Retail trade Consumer service	I, II. *Kind of Building* (general purpose for which building was constructed) such as: Housing Commercial Industrial
A, B. *Function Performed* (by each establishment as a whole) such as: Salesroom Personnel office Warehouse Retail store Separate lists under each kind of business, expressive of the purpose for which each establishment is set up.	A, B. *Type* (adapted to function) such as: Single-family house Multiple dwelling Professional offices Manufacturing lofts Separate lists under each kind of building, with many specialized building types but with vast majority of space listed under only few types.
1, 2. *Functional Differentiation* (within the establishment) Space required for: Offices Manufacturing Storage, shipping Selling A listing of space requirements, simple for highly specialized establishments, more complex for "general" establishments.	1, 2. *Description* (general form and size) such as Detached or row houses Elevator or walk-up Commercial, Office or Loft building Combinations of types will complicate this level.
a, b. *Measure of Space Required* (shape, access, services) Width, Depth, Ceiling height, Ground or upper floor, Frontage, Windows, Lighting, Heating, Construction, Floor loads, Finishes, Appearance.	a, b. *Measure of Space Available* Detailed description of space provided by the building, similar to the list at the same level for establishments.

SOURCE: John Rannels, *The Core of the City*, Columbia University Press, New York, 1956, p. 44. Reprinted by permission.

nodes for community interaction. A lack of sensitivity to the social functions served by such nodes may help to account for many of the shortcomings of less than successful city plans.

Urban symbolism

Over time certain places within a city become associated with certain types of activities. Streets acquire and keep functions and become known for them. The fame may spread widely until the street serves as a symbol for

the particular type of activity. Thus New York's Wall Street, Broadway, and Fifth Avenue and London's Bond Street and Fleet Street are associated with specific imagery even to people who have never seen them. The sociologist Anselm Strauss goes so far as to suggest that the city "can be viewed as a complex related set of symbolized areas."[17] Some areas may be known only to certain small groups, and each group may know only a few areas. But other places may be widely known because they are locations where the orbits of many separate social worlds intersect.

In European cities the downtown areas often have a rich symbolism attached to their old streets, squares, and buildings. So strong is the hold of this symbolism that many war-destroyed cities, such as Nuremburg and Warsaw, have deliberately recreated their new centers in the image of the old. The medieval street patterns are retained, and the heights and colors of the buildings controlled to recapture the flavor of the past. Although strong, the symbolism of the city center also has an illusive quality. Willem F. Heinemeyer studied the urban core of Amsterdam as a center of attraction.[18] He found that although many people liked to be there, it was not easy to explain exactly why. However, he did find that the areas perceived as particularly attractive were those with a diversity of separate functions that draw together a diverse public, creating what he terms a "forum quality" that further increases the attraction.

Urban symbolism has many facets.[19] The broadest level could include attitudes toward cities in general. Another level would be urban icons selected by individual urban areas as symbols of their city or the particular qualities seen as characteristic of the local atmosphere. Figure 5.3 illustrates some well-known urban icons that serve to represent their cities. Various portions of cities may also acquire symbolic meanings, such as the central business districts and neighborhoods, or suburbs, special areas, streets, or even individual elements. These images, once formed, tend to persist and often exert a great influence on people's behavior.

Attitudes toward the city

Many different themes have been detected in studies of attitudes toward cities. A broad survey by Morton and Lucia White indicated the essentially negative attitude toward the city that has been held by most of America's major writers and thinkers from Thomas Jefferson to Frank Lloyd Wright.[20] This contrasts with the often more positive tone toward small towns noted in the previous chapter. Although the reasons for dislike or fear or distrust of the city shifted, the negative view prevailed, and, according to the Whites:

> The American city has been thought by American intellectuals to be: too big, too noisy, too dusky, too dirty, too smelly, too commercial, too crowded, too full of

Boston

San Francisco

Seattle

New York

Washington, D.C.

St. Louis

Fig. 5.3 Landmarks that symbolize selected cities. (New York and Boston, American Airlines; St. Louis, Regional Commerce and Growth Association; Washington D.C., A Devaney, Inc.; Seattle, Washington State Department of Commerce and Economic Development; San Francisco, San Francisco Convention and Visitors Bureau.)

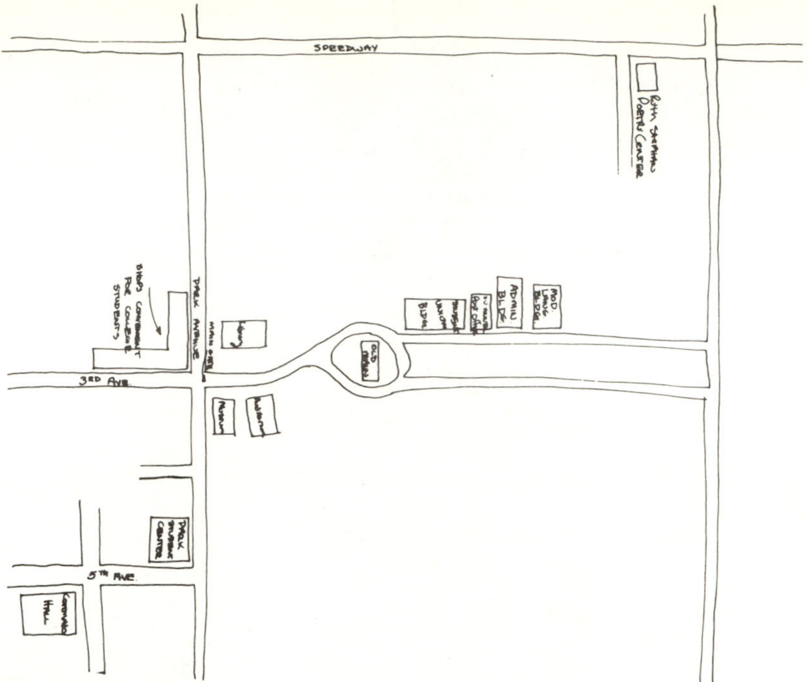

Fig. 5.6 English majors' map of University of Arizona campus.

samples of varying sizes. Derk DeJonge, in an early application of the technique, investigated the images of Amsterdam, Rotterdam, and The Hague as well as two residential neighborhoods in Delft and another in The Hague.[29] He found that formation of a map image is easiest where there is a street plan with a regular pattern and a single dominant path, characteristic nodes, and unique landmarks. Where the general pattern is not clear, more attention is given to isolated landmarks, individual paths, and visual details. In the neighborhood studies DeJonge found that difficulties in orientation may also arise where the structure is quite clear but the elements are too uniform to be distinguished. John Gulick, in the conclusions in his study of Tripoli, Lebanon, emphasized the role of sociocultural associations as well as visual clues in producing the urban image.[30]

Several studies I conducted in Tucson and in Chicago indicate the utility of the method at varying scales.[31] In each case it was clear that there are variations resulting from individual or group differences, but that in spite of these differences there is a high degree of conformity and consistency in the major elements. In one study, an areal sample of 200 students from 12 different departments at the University of Arizona were asked to draw maps of the university area. The individual maps varied considerably from

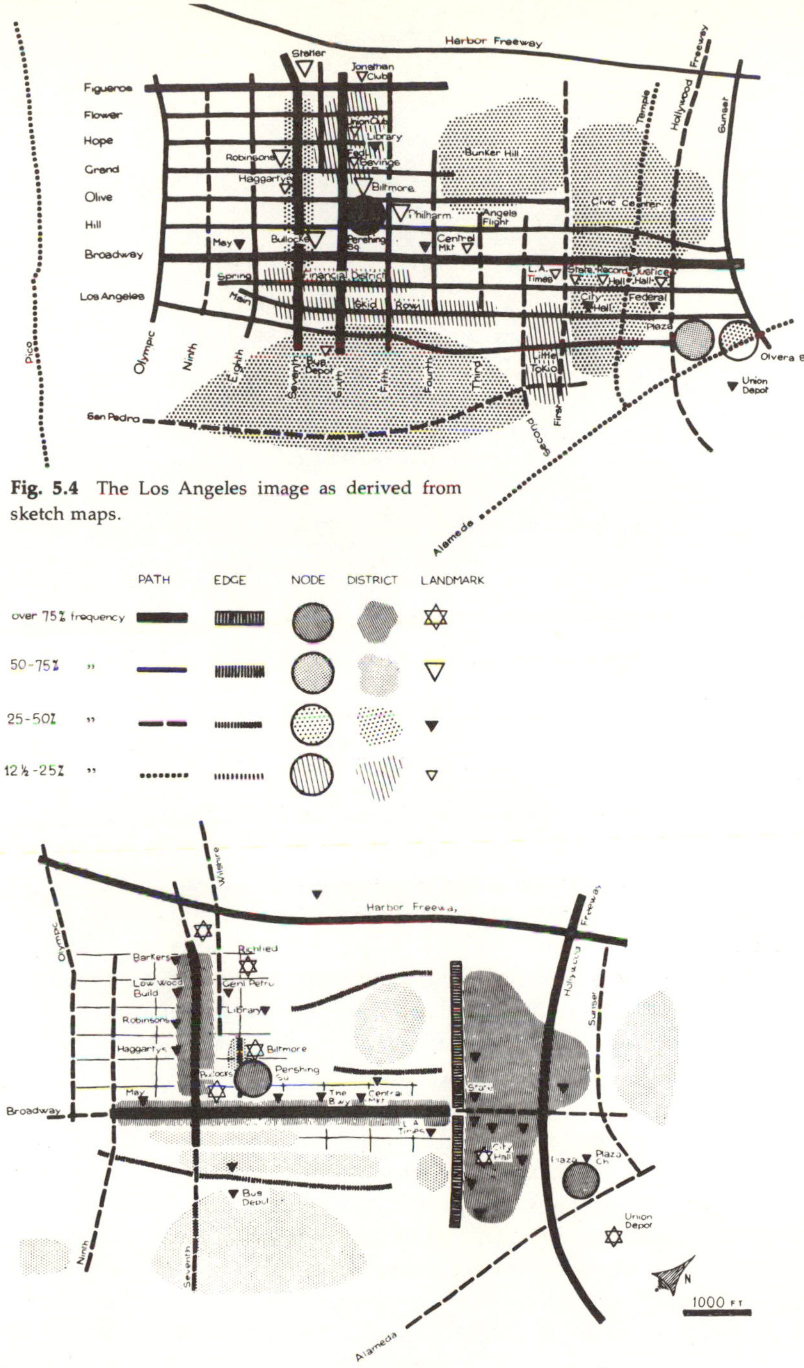

Fig. 5.4 The Los Angeles image as derived from sketch maps.

Fig. 5.5 The visual form of Los Angeles as seen in the field. (Figures 5.4 and 5.5 reprinted from the *The Image of the City* by Kevin Lynch by permission of The M.I.T. Press, Cambridge, Massachusetts. Copyright © 1960 by The Massachusetts Institute of Technology and the President and Fellows of Harvard College.)

systems of visual notation for urban areas.[28] The presence of certain landmarks or buildings on mental maps has led to questions of why these are known. The broader question of what elements people select from a complex reality to create their individual cognitive maps and how these develop has been investigated. Distortions in cognitive or mental maps have led to questions about subjective distance in cities.

In *The Image of the City* Lynch focused on the visual quality of the American city. He was concerned with the apparent legibility or clarity of the cityscape, or the ease with which its parts could be organized into a coherent whole. To get at this, he tried to find a means of measuring "public images," the common mental pictures carried by large numbers of the city's inhabitants. Although each individual probably has a different image, Lynch assumed that there would be broad areas of agreement, which might be expected to appear in the interaction of a single physical reality, a common culture, and a basic physiological nature. Residents were interviewed to determine the characteristics of the public image of the city. Those interviewed were asked to sketch a map of the city, provide detailed descriptions of a number of trips through the city, and list and briefly describe the parts they felt to be most distinctive. Using the data gathered in this way, Lynch was able to discuss the image of the city in terms of several main elements, paths, edges, districts, nodes, and landmarks. Putting them all together, he constructed a series of maps for the central areas of the three cities he investigated—Los Angeles, Jersey City, and Boston. The maps so constructed corresponded closely in most major features to the views of trained observers, indicating a consistency in the imageability of various areas or elements (compare Figures 5.4 and 5.5). However, the value of the interviews was underscored by the way the maps differed. The field analysis tended to neglect minor elements important for automobile circulation and some of the "invisible" effects of social prestige.

The great advantage of the draw-a-map portion of Lynch's technique is that it is essentially a projective test that allows for and demands a maximum of structuring by the subject. There are no clues given, as would be the case if the subject were asked to outline areas on a prepared map. In such a case the individual's ignorance of connection would be overlooked because the map would already provide the structure. The use of a blank sheet of paper means that the only available source for the map produced are the subjects' minds. They may not put on the map everything they know, but they cannot include anything they are not aware of. A major advantage of the method is that it can elicit responses that might be difficult to obtain by other means, such as the aforementioned invisible landscape. This is very important in attempting to assess experiences that people can rarely articulate.

With various modifications, Lynch's techniques have been applied in many studies, in different parts of the world, at different scales, and with

> immigrants, too full of Jews, to full of Irishmen, Italians, Poles, too industrial, too pushing, too mobile, too fast, too artificial, destructive of conversation, destructive of communication, too greedy, too capitalistic, too full of automobiles, too full of smog, too full of dust, too heartless, too intellectual, too scientific, insufficiently poetic, too lacking in manners, too mechanical, destructive of family, tribal and patriotic feeling.[21]

Legislation reflects the fact that American cities have been regarded negatively, as cities do not have an equal voice in the political order.[22] The European city fared somewhat better; for along with the idea of the city as vice, as Carl E. Schorske points out, European thought has contained other broad evaluations over the past couple of centuries—the city as virtue and the city beyond good and evil.[23] Even in the American city, however, some themes run counter to the general negative tone.

An excellent source for both positive and negative perspectives on the American city has been compiled by Anselm Strauss, who used as data a rich variety of historical and humanistic documents as well as the writings of social theorists. In addition to negative evaluations such as those quoted previously he documents such positive perspectives as the city as a place of diversity, as a place of fun and adventure, and as a place of social opportunity. The contrasting viewpoints of scholars and novelists are noted in one essay in which Strauss asserts:

> Taken as a group, novelists have portrayed life in the city not only more dramatically—more humanly, if you wish—than their scholarly contemporaries, the sociologists, the geographers, the planners; but they have been less heir perhaps to inherited intellectual views which divert gaze and cramp vision.[24]

Novelists perceive the city as animate and potent, capable of "making or breaking" people. Their varied perspectives provide a fertile source of data on environmental perception and behavior at all scales. Novels are particularly valuable for their personalized accounts of people-environment interactions in specific areas. The deep insights into the psychological and symbolic aspects of the people-environment relationship garnered in novels could well serve as hypotheses for more rigorous scientific studies.

The image of the city

The Image of the City by Kevin Lynch is probably the best-known work in the field of environmental perception.[25] It is also one of the earliest, having been published in 1960.[26] The stimulus value of the approach developed by Lynch is attested to by the large number of direct applications of all or portions of his methodology and by the degree to which various elements have been developed on their own.[27] Lynch's system of describing the city has helped stimulate new efforts among designers to develop broad-scale

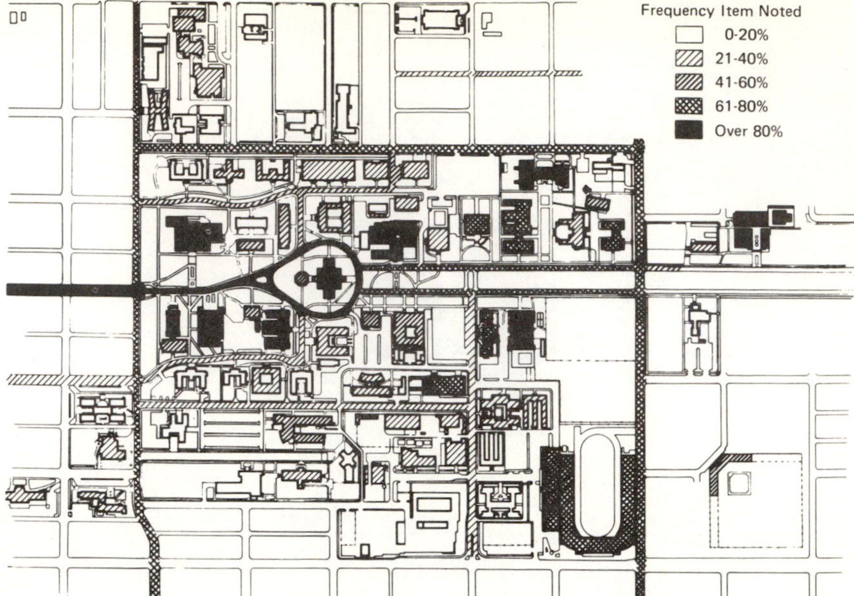

Fig. 5.7 Image of University of Arizona: Women's physical education class.

almost complete campus charts to simple diagrams of the student's main paths, as in Figure 5.6. Here the student, an English major, indicated the route from her dormitory to the services considered essential for her activities. At later stages one might expect a more complete image. The composite map for each of the departments varied in small ways, mainly by including with greater frequency the buildings where those students' classes were held and others nearby. An example is provided of the composite map for the women's physical education students (Figure 5.7). Their image was much more elongated than most. They generally included the women's physical education building at the extreme eastern edge of the campus as well as most of the other buildings along the main mall, which they walked along to get there. But, like all the other departmental maps, theirs tended to emphasize the same central area of high imageability and ignored elements beyond the campus proper. In addition, all composite maps tended to show a gradually fading image outward from the strong central core. Such agreement is more likely in a small, relatively uniform area like a campus.

The effect of group differences in a more diverse area can be seen in a study of the Chicago Loop. In this case many of the differences stem from degree of familiarity and in terms of how the Loop is entered and for what purpose. Three groups were tested; each drew maps and listed the most distinctive areas in the Loop. All the groups noted certain areas and elements and omitted others, as shown in Figures 5.8 to 5.10. All the groups

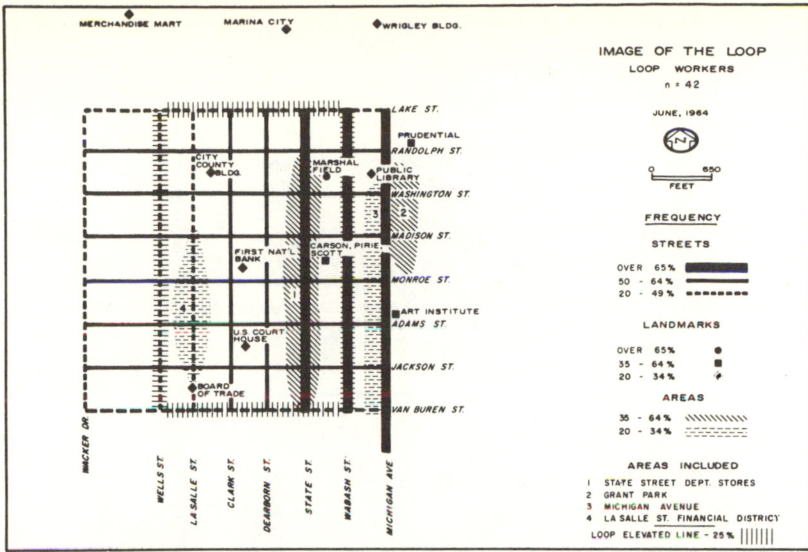

Fig. 5.8 Image of the Loop: Loop workers.

noted the grid pattern and straightened out any variations from it. In all cases the degree of clarity of the image diminished westward and the north-south streets were dominant. State Street with its department stores, the contrast between the open space of Grant Park and the massive wall of Michigan Avenue, canyonlike LaSalle Street dominated by the towering Board of Trade Building, and the clean sweep of the north branch of the Chicago River lined by major landmarks were major images that appeared to some degree in each composite map. But there were differences: The workers within the Loop tended to have a somewhat more tightly defined area with more internal detail, while the groups from outside tended to include a broader area with much more emphasis on external landmarks.

Although many of the Loop workers delimited the Loop in terms of the elevated railroad line, they often included streets and landmarks beyond this boundary. Figure 5.11 shows one such map drawn by an employee of the Board of Trade. In many instances surprisingly parochial views appeared, as only the place of employment and a few nearby or closely associated businesses were included. An extreme example is provided by the map of a State Street department store clerk, in which State Street becomes the Loop (Figure 5.12). But the parochial tendency was general. Retail workers tended to include other retail shops, government employees included government buildings, and most of the maps tended to include more detail adjacent to the map drawer's place of employment.

Both groups from outside included Lake Shore Drive, the Congress Expressway, Lake Michigan, and the north branch of the Chicago River. The students from Crane Teachers College, mainly residents of the west

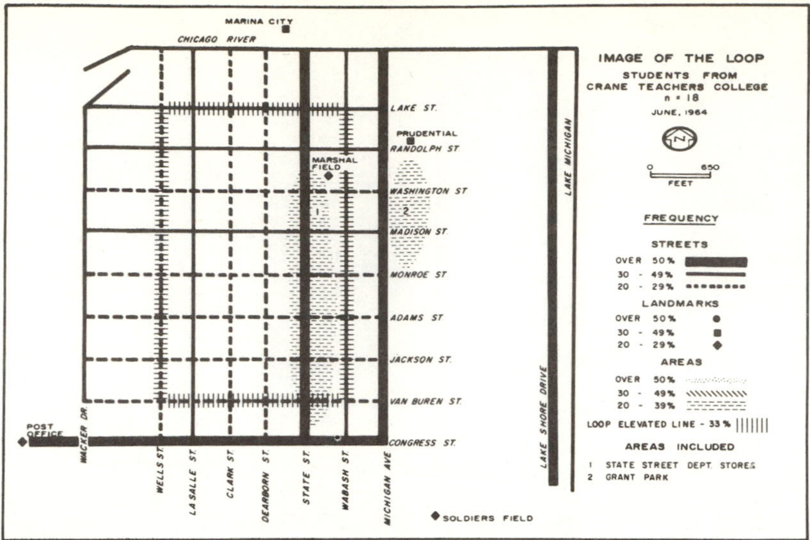

Fig. 5.9 Image of the Loop: Students from Crane Teachers College.

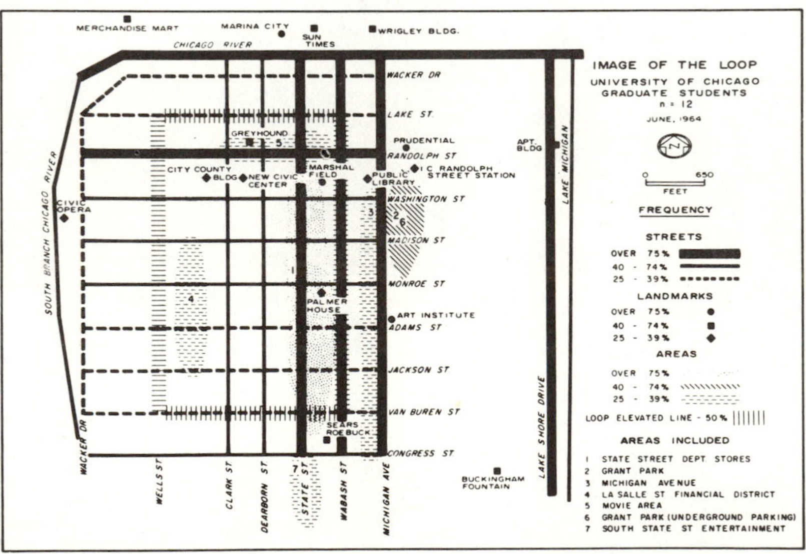

Fig. 5.10 Image of the Loop: University of Chicago graduate students.

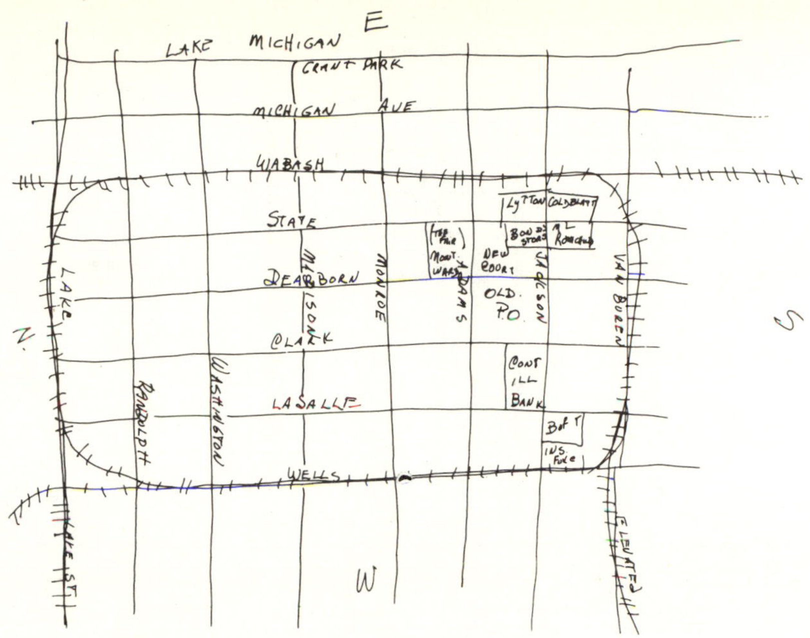

Fig. 5.11 Image of the Loop: Individual map showing elevated railroad line.

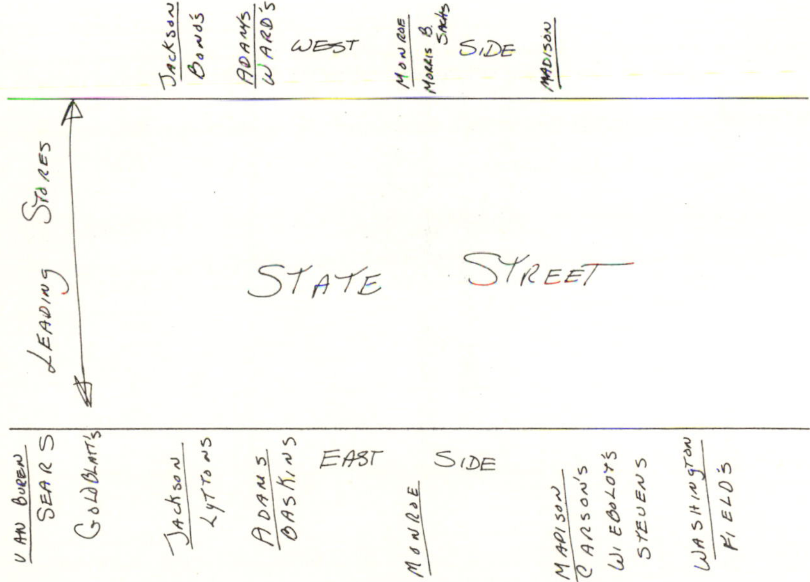

Fig. 5.12 Image of the Loop: Individual map by department store clerk.

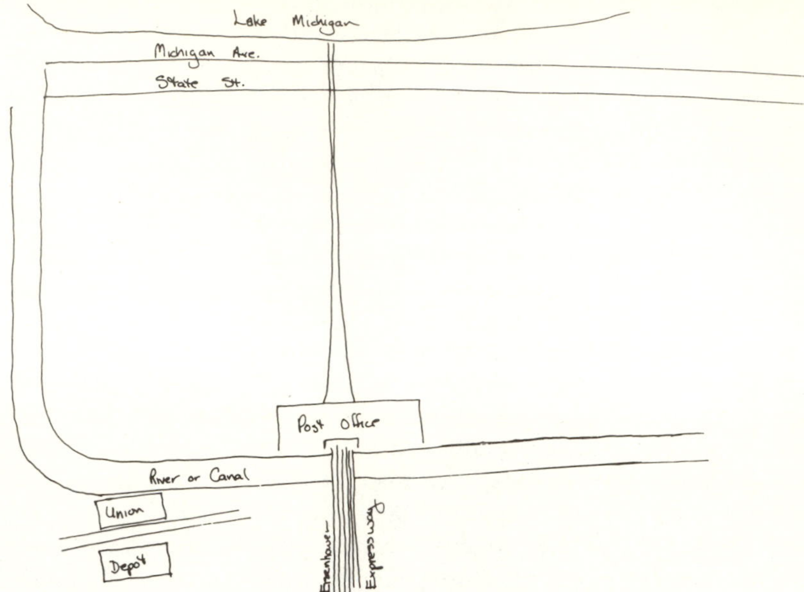

Fig. 5.13 Image of the Loop: Individual map illustrating dramatic entry under post office.

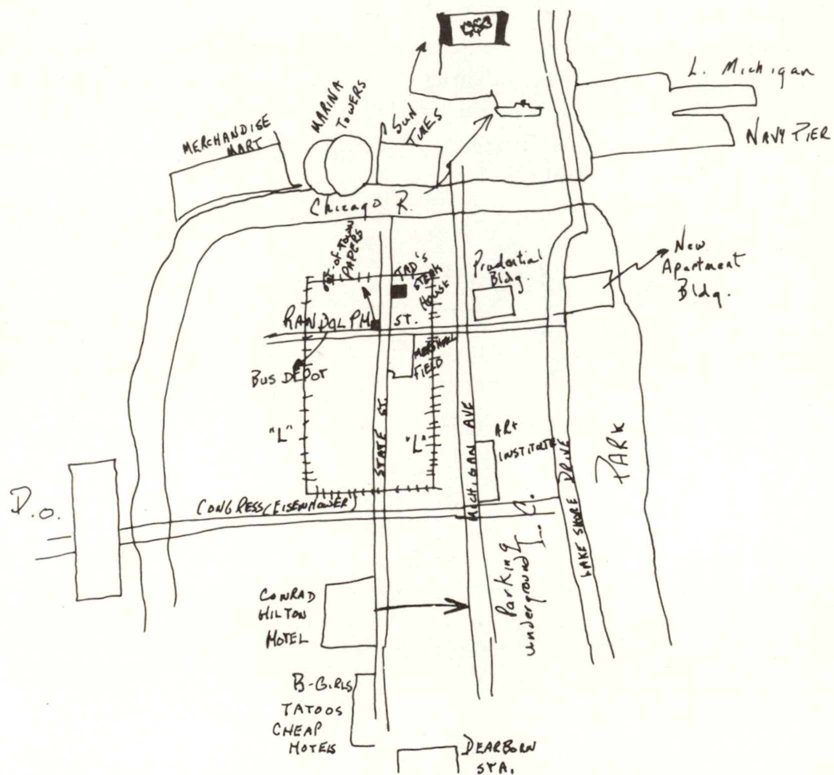

Fig. 5.14 Image of the Loop: Individual map of Chicago graduate student.

and southwest suburbs, also included the Post Office, which provides a dramatic entry as the expressway from the west sweeps under the building (Figure 5.13). The students from the University of Chicago included the Randolph Street Station and Randolph Street with greater frequency than the others, no doubt because they often enter the Loop at this point after arriving from the south side on the Illinois Central. They were all single, from out of town, and, as their image indicates, interested in the tourist and entertainment attractions offered by the Loop. Figure 5.14 provides an individual example by a University of Chicago student from Canada who includes a Canadian ship delivering newsprint to the Sun-Times Building.

Probably the most massive collection of data of this type is Hans-Joachim Klein's delimitation of the town center in Karlsruhe, which was based on 1,118 interviews.[32] He used a pack of cards with places and roads marked on them. The subject picked out the ones he considered to be in the town center. Although there was a coherent central area consistently included, many interesting variations appeared in the extent of the area. Older residents included a broader area than newcomers. Male-female and socioeconomic differences appeared, and place of residence was significant. Those living within the center had a narrower image than those farther away. The city image of people in the western quarters of the city differed from that of people in the east, although images overlapped somewhat, as may be seen in Figure 5.15. The systematic differences described by Klein are similar to the patterns found in relation to the Chicago Loop and in later studies of downtown Washington[33] and the Amsterdam central business district.[34]

The studies to date indicate the flexibility of Lynch's technique, which has been applied in many different areas, at different scales, and in relation to several different problems. Robert W. Kates, in a study published in 1970, tabulated close to thirty such image studies known at the time to

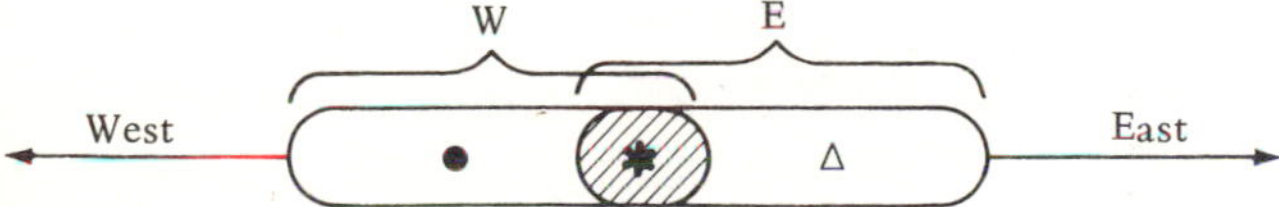

W – City-image of people living in western quarters
E – City-image of people living in eastern quarters
Δ – Earlier geographic focus of the town
✱ – Actual focus of the city
⊘ – City-region in the image of all inhabitants
● – Average imaginative focus of the city (provided a paramount westward growth of the total town)

Fig. 5.15 Variations in city image of Karlsruhe residents. (Hans-Joachim Klein, "The Delimitation of the Town-centre in the Image of its Citizens," in *Urban core and Inner City*, ed. E. J. Brill, Leiden, Netherlands, 1967, pp. 286-306. Reprinted by permission.)

Kevin Lynch and himself.[35] They had been carried out in such varied cities as Boston, Berlin, Bangkok, Stockholm, Rome, and Rye (New York). But even at that time the number was not complete, and many more have since appeared.[36]

For those engaged in city planning and design, the Lynch technique provides a starting point. By systematically studying images, it is possible to see which parts of the city appear clear, orderly, or memorable, and how the parts fit together as a whole. Where the image is unclear, it seems likely that problems of disorientation will arise. These can be corrected by better design of the paths, edges, nodes, districts, and landmarks that make up the city. It is assumed that a city with a clear, coherent image is one that is a pleasure to live in. There is no distress from disorientation, as one always understands one's relationship to the larger whole. Energies can thus be spent in more positive activities than finding one's way.

A city with a coherent image

In *Fleeting Glimpses,* Denis Wood described a Mexican city that is loved by its inhabitants and possesses an exceptionally clear image.[37] His study includes, in addition to the usual visual image of the Lynch technique, information based on color, smell, sound, and temporal differences. San Cristobal emerges as a city with an exceptionally clear image and as an exciting city to live in. A major factor in the coherence of the image is what Wood calls "replication," or the recreation of similar forms on varying scales. This is illustrated in Figure 5.16, where the spatial form of the house is seen as similar to the form of the barrio or neighborhood, which in turn is repeated at the city scale. Thus each home has a central patio surrounded by rooms which corresponds to the barrio plaza surrounded by individual houses and the main plaza or *zocalo* surrounded by the separate neighborhoods. The patio serves as the main recreational area for children, the barrio plaza, for young adolescents, and the *zocalo,* for older adolescents and adults. Politics, religion, and fiestas are similarly evident on all levels, so that skills learned at the lowest level are easily transferable to each succeeding level of the hierarchy. The barrio, or neighborhood, is a critical intermediate level in this process of replication that binds the home to the city. It is one that Wood emphasizes is lacking in Anglo-American cities. He suggests that his results be correlated with information concerning satisfaction, health, spiritual well-being, and so on to investigate the relationship between the coherence of the city image and the physical and spiritual health of the city inhabitants.

Ciudad Guayana image studies

Some important recent extensions to the understanding of urban perception stem from the work of Donald Appleyard with the Ciudad Guayana

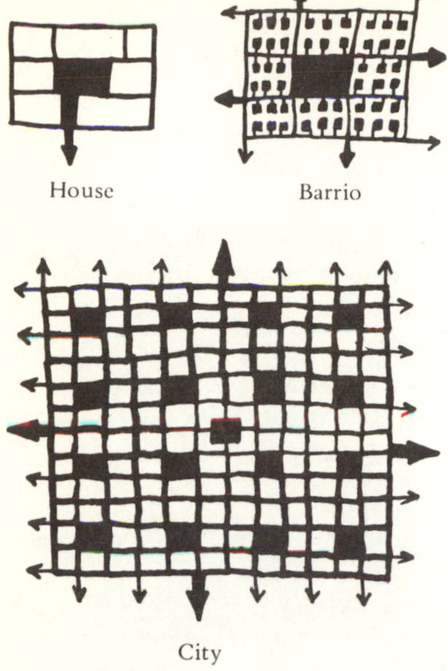

Fig. 5.16 Idealized city plan showing the individual home, the barrio, and the city. (Denis Wood, *Fleeting Glimpses*, Clark University Cartographic Laboratory, Worcester, Massachusetts, 1971, p. 223. Reprinted by permission.)

project in Venezuela.[38] A team of twelve interviewers under the supervision of a sociologist interviewed a total sample of about 320 subjects drawn from each of the distinctive environments in the existing city. The interviews were paralleled by systematic environmental surveys by trained observers. The study marked the first time such a survey was conducted on this scale as part of the design of a new city. The results are fascinating and full of ideas that challenge the ingenuity of the planner who wishes to consider the needs, purposes, and abilities of all population groups in city design. The great gap between the perceptions of planners and the local people illustrated by Appleyard has already been noted. Here we will deal only briefly with two other aspects: first, why buildings are known,[39] and, second, some styles and methods of structuring a city.[40]

Appleyard investigated why certain elements were known and included in an inhabitant's mental representation of Ciudad Guayana. All buildings, establishments, and other landmarks recalled by the inhabitants were recorded, photographed, and scaled according to an array of attributes considered critical to their identification and recall. It was assumed that an inhabitant would include a building or place for some combination

of four reasons: (1) the distinctiveness of its physical form; (2) its visibility in the cityscape; (3) its role as a setting for personal activities, use, and other behavior; (4) its perceived cultural significance. The scaling for form intensity is shown in Tables 5.2 and 5.3, which indicate the verbal and graphic rules.

Visibility and significance scales were also developed, and each separate building attribute was correlated with recall frequencies in map drawing, verbal recall, and trip recall. Thus Appleyard was able to provide precise statements on the relative significance of each of the attributes in terms of building recall. For example, movement and contour were the most powerful identifying qualities in map responses, whereas size and shape rose in importance in verbal recall. Movement, contour, size, shape, and surface were more important form attributes than quality or signs. Visibility was important. Buildings at decision points in the city, such as intersections, bus stops, bends, and ferry crossings, were recalled frequently even if otherwise insignificant. Similarly, those closest to the line of vision along the main paths were more likely to be remembered. The role of use was seen as important, indicating that many inhabitants were looking at the environment as a setting for activity. Putting it all together, Appleyard asserts "the archetypal patterns of a citizen's urban knowledge take the form of three concentric zones: the used, the visible, and the hearsay world beyond."[41]

Appleyard suggests certain ways the planner and designer may use his vocabulary of urban attributes. By coordinating urban form, visibility, and action with community significance, a more meaningful city could be created. If the community could decide which buildings should be known by the population, "it should be possible to adjust the attributes of form and visibility to enable each building to achieve the desired level of recognition."[42]

Another major part of Appleyard's work in Cuidad Guayana was concerned with styles and methods of structuring a city. In classifying the inhabitants' sketch maps of Cuidad Guayana, he found that they varied according to the type of element predominantly used, and according to their level of accuracy. Two main types of elements were used: sequential elements, mainly roads and nodes, and spatial elements, which included individual buildings, landmarks, and districts. Within each of these two types, four subtypes were identified, which are illustrated in Figure 5.17. One can see in the sequential types gradual improvement in quality from fragments through chains, branch, and loop to the more complex, network maps. A less clear gradation was apparent in the spatial types, which included scattered, mosaic, linked, and patterned maps. Clearly, people use a great variety of methods to conceptualize the city. However, sequential maps were most common, comprising three quarters of all the maps sketched. Appleyard states:

Table 5.2 Form Intensity Scales (Verbal Rules)

	Low	**Medium**	**High**
Movement	No movement.	Potential movement, parked cars, few people.	Many people, moving cars, flags waving, water falling.
Contour	Slurred boundaries hidden by vegetation attached to other houses.	Semidetached corner buildings.	Isolated buildings with sharp contours.
Size	Single-storey buildings: houses.	Two-storey buildings: movie houses.	Over two-storey buildings: industrial sheds, steel mill, General Electric.
Shape	Simple.	Two or three block buildings.	Complex building divided into several parts.
Surface	Plain white.	Colored.	Brightly contrasted colors and textures.
Quality	Bahareque (wattle), mud floors outside, no fences.	Modest materials, walls, garden.	Landscaped, fenced, expensive materials, clean conditions.
Signs	No signs.	Small signs.	Large signs readable from a distance.

SOURCE: "Why Buildings are Known: A Predictive Tool for Architects and Planners," by Donald Appleyard. Reprinted from *Environment and Behavior*, Vol. 1, No. 2 (December, 1969), 134-135. Reprinted by permission of the publisher, Sage Publications, Inc.

Table 5.3 Form Intensity Scales (Graphic Rules)

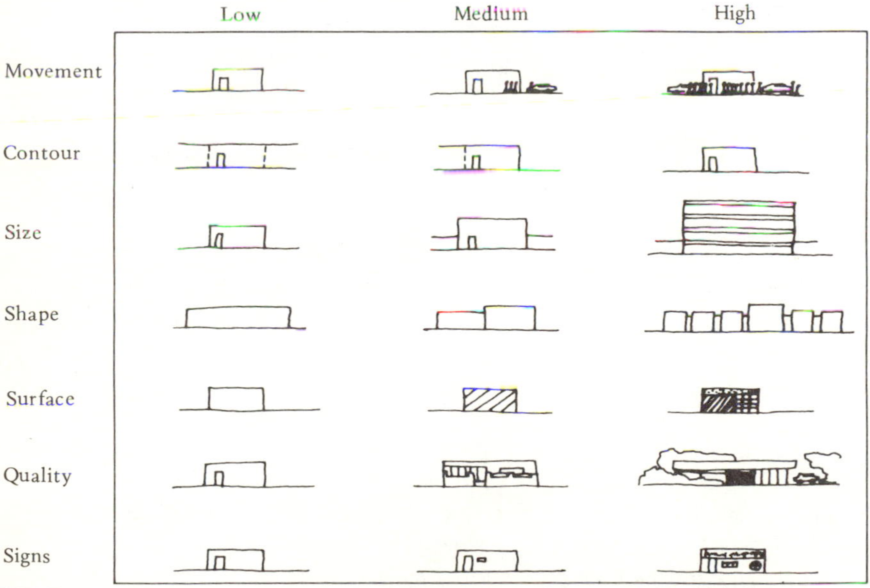

	Low	Medium	High
Movement			
Contour			
Size			
Shape			
Surface			
Quality			
Signs			

SOURCE: "Why Buildings are Known: A Predictive Tool for Architects and Planners," by Donald Appleyard. Reprinted from *Environment and Behavior*, 1 (December, 1969) 134-135. Reprinted by permission of the publisher, Sage Publications, Inc.

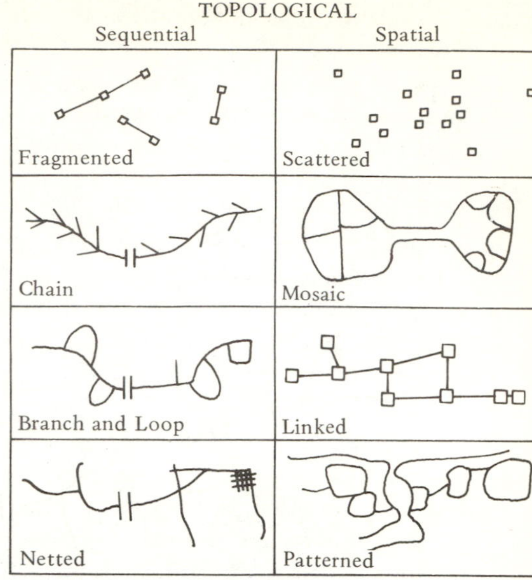

Fig 5.17 Methods of structuring maps. (Donald Appleyard, "City Designers and the Pluralistic City." Adapted from *Planning Urban Growth and Regional Development*, p. 437, by Lloyd Rodwin, ed., by permission of The M.I.T. Press, Cambridge, Massachusetts. Copyright © 1969 by The Massachusetts Institute of Technology.)

In the cognitive representations of large cities, people have to schematize drastically if they are to gain any overall comprehension of urban structure. They extract dominant reference points, a group of districts, or a single line of movement on which to hang their recollections.[43]

Many group differences in map construction were found. The less educated people were less able to conceptualize, and often the interview situation marked the first time they were faced with the task of drawing a map. They were more error-prone, and their maps "conveyed a strong sense of subjective experience."[44] The more educated people were able to approach the task more objectively and fit their maps together more coherently through use of inferential structuring. In other words, they generalized from experience of environmental relationships to infer aspects of present patterns. Interestingly, such inferences did not always lead to accuracy, for images derived from direct experience became confused with those based on interpretations of a given experience.

In Ciudad Guayana about half the sample used buses as their dominant means of transportation, while one quarter used cars or collective taxis. This had a profound influence on the type of map drawn. The group who

traveled by car were able to sketch a coherent and continuous system, while 80 percent of those who used only the bus were unable to do so.

Spatial elements were more common on the maps of the individual's local area than for other parts of the city. It was not simply a case of a "spatially well-developed home area with strands of sequential knowledge stretching out in other areas of the city," for "well-structured islands of knowledge"[45] of other parts of the city were common as well. Use of spatial elements became more common with increasing familiarity. Newcomers produced predominantly sequential sketches, and their maps were much more restricted. However, after about six months to a year, they had fewer errors per zone in the parts of the city they could draw than longer-time inhabitants, indicating a high level of interest and concern.

The major factors assessed by Appleyard to be important in explaining the different modes of structuring maps by various population groups were cognitive differences, travel mode, and familiarity. Our knowledge of the way people develop mental maps for orientation and understanding of urban areas can be advanced by more of this type of work, which fits within the framework for further research advocated by Roger Downs and David Stea.

Cognitive maps

Downs and Stea have stressed the importance of the image in relating environmental behavior to perception.[46] They call for a research focus on these images, mental maps, conceptual spaces, schemata, or cognitive maps to explain environmental behavior. To explore the questions they consider important in advancing knowledge of our cognitive representation of our spatial environment, they suggest three major research focuses. They are, respectively: the elements that comprise cognitive representations, the relations between elements, and the surfaces resulting from the relationships between elements.[47, 48]

Downs demonstrated the cognitive complexity of a single spatial element in his investigation of an urban shopping center.[49] Since the image of a spatial element is often assumed to exert some effect on spatial behavior, it seems important to determine the cognitive categories or dimensions of this image. For the shopping center, this would include categories concerned with the nature of the center and others dealing with the beliefs about attributes of the center and the relative importance of these attributes.

Downs used the semantic differential technique, employing the 36 pairs of adjectival phrases in Table 5.4. He interviewed over 200 women in Bristol, England. Each was asked to evaluate the shopping center in terms of each pair of phrases. They indicated on a seven-point scale which pair was closest to describing the shopping center. In this manner a rating of

Table 5.4 Nine Hypothesized Cognitive Categories and 36 Semantic Differential Scales

(1)	**Price**	
	1 competitive	uncompetitive
	2 many bargains	few bargains
	3 good value for money	poor value for money
	4 many price cuts	few price cuts
(2)	**Structure and design**	
	5 well designed	badly designed
	6 simple layout	complicated layout
	7 designed with shoppers in mind	not designed with shoppers in mind
	8 wide pavements	narrow pavements
(3)	**Ease of internal movement and parking**	
	9 easy to cross roads	difficult to cross roads
	10 easy to park	difficult to park
	11 not congested	congested
	12 easy to walk around in	difficult to walk around in
(4)	**Visual appearance**	
	13 well-kept shops	badly kept shops
	14 tidy	untidy
	15 clean	dirty
	16 attractive	unattractive
(5)	**Reputation**	
	17 good reputation	bad reputation
	18 generally well known	generally little known
	19 generally popular	generally unpopular
	20 recommend to friends	wouldn't recommend to friends
(6)	**Range of goods**	
	21 good choice	poor choice
	22 wide range	narrow range
	23 well stocked	badly stocked
	24 can get it	can't get it
(7)	**Service**	
	25 helpful	unhelpful
	26 friendly service	unfriendly service
	27 good service	poor service
	28 polite	rude
(8)	**Shopping hours**	
	29 late closing	early closing
	30 convenient opening times	inconvenient opening times
	31 good for evening shopping	bad for evening shopping
	32 always somewhere open	never anywhere open
(9)	**Atmosphere**	
	33 busy	not busy
	34 relaxed atmosphere	tense atmosphere
	35 personal	impersonal
	36 friendly atmosphere	unfriendly atmosphere

SOURCE: "The Cognitive Structure of an Urban Shopping Center," by Roger Downs. Reprinted from *Environment and Behavior*, Vol. 2, No. 1 (June, 1970), p. 22 by permission of the publishers, Sage Publications, Inc.

each pair was obtained for each respondent. The technique was considered appropriate, as it provided a structure for eliciting reactions to the shopping center that might be difficult to express directly.

The ratings of all the respondents were factor-analyzed to explore the group image of the downtown shopping center in Bristol. Eight cognitive categories were abstracted. In descending order of importance, they are: quality of service, price, structure and design, shopping hours, internal pedestrian movement, shop range and quality, visual appearance, and traffic conditions. These could be grouped into two types, retail establishment factors and factors related to the structure and function of the shopping center. Downs' study indicates that "the image is a complex construct, and that there are significant interrelations between spatial cognition, urban form, and human behavior of which we are almost completely ignorant."[50]

Cognitive distance

One aspect of the relation between elements is how distances are perceived in an urban area. Apparently, there are major distortions depending on the direction. For example, in Dundee, Scotland, Terence Lee found that, in estimating walking distances, students consistently estimated the distances in the downtown direction as shorter than those away from the center.[51] He suggests that the satisfactions of the center build up a focal orientation that causes a foreshortening of perceived distances in the inward direction. Curiously, another study in Columbus, Ohio, found an opposite effect.[52] Underestimation of distance was the rule in areas away from the central business district (CBD), while overestimation was common in the area toward the CBD. The authors interpret the increase in perceived distances as the result of a denser packing of land uses toward the CBD, together with increased congestion and travel time. It seems likely that the discrepancies between the studies cited could be explained by travel mode, as the English students were estimating walking distances while their American counterparts were probably estimating driving distances.

Several other experiments related to subjective distance were carried out in the geography department of the Ohio State University. These are reviewed in detail by Reginald Golledge and Georgia Zannaras.[53] They suggest that familiar locations were perceived closer than less familiar points an equal distance away. On reconsidering the data regarding the perceived greater distances in the downtown direction, they found that this held true for small distances. As the estimated distances became larger, there was a greater degree of convergence between objective road distance and cognitive distance. Much more research is required to sort out all the variables and their implications. Such research is potentially important, as cognitive distance may explain many types of spatial behavior better than real distances.

Ages and stages in the development of cognitive maps

A major question is that of the ages and stages at which the various cognitive skills required to construct mental maps are developed. There have been, for example, a few attempts to use the Lynch map-drawing technique among people of different ages to see how the city image develops and changes over time.[54] An excellent review by Roger Hart and Gary Moore indicates how recent work on the development of spatial cognition relates to the fundamental philosophical concepts of space and the major psychological theories of cognitive development, such as that of Piaget.[55] Their review indicates the present knowledge of the stages people pass through from infancy to adulthood in the general progression from concrete to abstract representation and thought about space. A grounding in such a broad theoretical framework would provide a firm foundation for further research on cognitive maps at all scales.

Another different approach is being pursued by the Place Perception Project at Clark University.[56] An example of an early investigation in this project is the study by James M. Blaut, F. McCleary, Jr., and America S. Blaut, which tested whether North American and Puerto Rican children of school-entering age could interpret and utilize an environmental map without prior training.[57] They presented the six-year-olds with vertical, aerial photographs and asked them to identify features. To do so requires "the cognitive leap to an imaginary vantage which is rotated about 90° from the normal and reduced in scale several thousand times."[58] Yet most of the children performed the feat with no difficulty. The authors suggest that children develop such abilities very early through playing with toys that represent models of the real world.

> . . .toys and toy-like objects stand for entities at the perceptual scale, the floor or ground stands for the surface of the world, and the aggregation of toy-abstract into macro-environmental forms—reduced and rotated—is a means of accomplishing in play the macro-environmental learning which cannot be accomplished in reality. Perhaps it is also reasonable to suspect that this activity is one of the earliest uses of model-thinking and, therefore, of scientific method.[59]

The major aim of this research was to determine how children proceed from immediately perceived experience to concepts of place, so that, with a better understanding of the ages and stages involved, better teaching methods could be developed. Other researchers have been more concerned with the planning implications of what people attend to at various ages and stages and how the city can be designed to aid in the process of healthy human development physically, mentally, and spiritually.[60]

The previous sections on the image of the city were basically concerned with the total physical pattern of the city. The discussion illustrated how many individual activity patterns may be combined to form group images

of the environment. An underlying idea was that through a better understanding of the cognitive maps of city dwellers, new insights might be gained that would be useful for broad-scale city design. The city envisaged would be designed to support of human needs and purposes. From specific studies of images of various cities, there was progression to the more abstract issue of cognitive maps in general, and we briefly noted a few examples of aspects of these maps. From the abstractness of cognitive maps and the comprehensive city view of image maps we will now turn to the more concrete as we examine the psychological atmosphere of cities. The concern shifts from emphasis on spatial patterns to a focus on individual or social reactions to the complex array of stimuli that surround us in the city.

The metropolitan personality

The hectic pace of city life contrasts strikingly with the slower, more regular rhythms of rural areas and small towns. Associated with the contrasts in types and amount of stimulation are personality adjustments. William Michelson reviews some of the earlier arguments, which he says typically stressed the pathological effects on the residents of the physical aspects of the city.[61] Thus, for example, James S. Plant, describing mainly juvenile residents in the crowded urban slums, emphasizes the "hardening" process, the destruction of illusions, feelings of insecurity and inferiority, and a general characteristic of "mental strain."[62] Georg Simmel speaks of the development of the blasé attitude, a blunting of discrimination so that "the meaning and differing value of things, and thereby the things themselves, are experienced as insubstantial."[63] Man in the metropolis "reacts with his head instead of his heart" and becomes more calculating, punctual, and exact. He behaves toward other people with reserve, which makes him appear cold and heartless in the eyes of small-town people. Louis Wirth stressed the interrelationships between large size, high density of population, and heterogeneity of inhabitants as essential characteristics of urbanism.[64] He characterized the contacts in the city as impersonal, superficial, transitory, and segmental and suggested that "the reserve, the indifference, and the blasé outlook which urbanites manifest in their relationships may thus be regarded as devices for immunizing themselves against the personal claims and expectations of others."[65] He also pointed out that the frequent close physical contact coupled with great social distance could give rise to loneliness and that people are regarded as categories rather than as individuals. The summation of all these largely negative adjustments indicates that urban existence poses great problems for the individual personality. However, there are also great advantages to city life that should be mentioned as well.

Simmel stated that the psychological basis for the metropolitan personality is the "intensification of nervous stimulation" to which one is con-

stantly exposed. A key question that must be answered if we are to understand modern life is "how the personality accommodates itself in the adjustment to external forces"; or, how "the person resists to being leveled down and worn out by a social-technological mechanism."[66] In addressing himself to this question, Simmel noted positive as well as negative consequences. The intense stimulation that may lead some people to adopt the blasé attitude may stimulate other individuals to achieve their peak potential. The impersonality of urban life may result in a heightened individuality or merely an elaboration of individuality, as in the adoption of extravagant mannerisms in order to stand out in a striking manner by being different. Associated with the reserve of urban dwellers is a kind and amount of personal freedom that contrasts with "the pettiness and prejudices which hem in the small town man."[67] But a more important freedom derives from the cosmopolitan nature of the city.

> For it is the decisive nature of the metropolis that its inner life overflows by waves into a far-flung national or international arena. . . The most significant characteristic of the metropolis is this functional extension beyond its physical boundaries. And this efficacy reacts in turn and gives weight, importance, and responsibility to metropolitan life.[68]

Discussing what the city dweller or the city needs, various authors may combine in different ways the advantages and disadvantages of the small town and the large city. It would be nice if it were possible to combine the advantages of each without any of the disadvantages. Thus one could have the warmth, spontaneity, and security of small-town life combined with the variety, excitement, and great intellectual stimulation of the city. But the characteristics of the ideal city and the methods of creating it vary considerably with the author. Jane Jacobs, for example, criticizes orthodox city planning for destruction of urban vitality as older areas are bulldozed to be replaced by bleak, monotonous, dehumanized high-rise slabs.[69] She describes, in a masterful fashion, the intricate web of human relationships fostered by certain intimate urban neighborhoods. In such areas people mind their own business, and yet all cooperate in various ways to create a pleasant atmosphere with a sense of belonging and shared civic responsibility. Jacobs states that people desire diversity, that when this is destroyed urban vitality ebbs and no longer are there "eyes on the street" to insure the safety of the local urban residents. She advocates concentration of people; short blocks, a diverse array of buildings in terms of age and condition, and districts with many different functions to insure the presence of people on the streets at all hours.

Although many planners admire Jane Jacobs' fresh approach and excellent descriptions of urban neighborhood life, they do not always agree with all her prescriptions. Herbert Gans, for example, suggests that the ethnic neighborhoods she often cites as good examples of urban vitality are homogeneous in terms of people and buildings.[70] The lively street life is

not derived from diversity but from their working-class culture. Furthermore, he argues that middle-class people would not want to raise their children in either the working-class or the Bohemian neighborhoods described by Jacobs as desirable. Lewis Mumford asserts that

> . . . one solitary walk through Harlem should have made Mrs. Jacobs revise her notions of the benefits of high density, pedestrian filled streets, crosslines of circulation, and a mixture of primary economic activities on every residence block, for all these ideal conditions are fulfilled in Harlem—without achieving the favorable results she expects of her prescription.[71]

Mumford argues that the city by its very size has got out of hand and that further encouragement of uncontrolled growth is destructive. Instead, he suggests that

> . . . one begins with the smallest units and restores life and initiative to them—to the person as a responsible human being, to the neighborhood as the primary organ not merely of social life but of moral behavior, and finally to the city, as an organic embodiment of the common life, in ecological balance with other cities, big and small, within the larger region in which they lie.[72]

Regardless of the theory of the ideal urban habitat, one thing is certain: To provide it, we must learn much more about the psychological dimensions of urban life. In spite of the early appearance of statements such as those noted previously, there is still very little objective evidence on which to judge their adequacy. Some fascinating, recent attempts to develop methods of measuring psychological dimensions of the city are described here. More of this sort of research is needed to fully explore the experience of living in cities.

Urban behavior as adaptations to overload

Stanley Milgram, a social psychologist at the City University of New York, interprets urban behavior in light of the concept of "overload."

> This term, drawn from systems analysis, refers to a system's inability to process inputs from the environment because there are too many inputs for the system to cope with, or because successive inputs come so fast that input A cannot be processed when input B is presented.[73]

The kind of behavior observed in cities is seen as a series of adaptations to "overload." The individual is forced to set priorities and make choices. Low-priority inputs are disregarded, and the intensity of inputs is diminished, so that only weak and superficial forms of involvement with others occur.

The ultimate adaptation to an overloaded social environment is to totally disregard the needs, interests, and demands of those whom one does not define as relevant to the satisfaction of personal needs. . .[74]

The effects of adaptations to "overload" may be seen in all types of behavior, from bystander intervention in crisis to willingness to trust and assist strangers or to observance of everyday civilities. Milgram describes many simple but ingenious naturalistic experiments designed to test some of the differences between town and city behavior. One example was a test of the degree to which city and town dwellers were willing to trust and assist strangers. This was measured by the willingness of homeowners to allow strangers to enter their homes to use the telephone. Student investigators rang doorbells, explained that they had lost the address of a friend nearby, and asked to use the phone. The female experimenters had better success in gaining entry than the male ones, but in all cases the students did at least twice as well in towns as in cities. Furthermore, the perceived danger level of living in Manhattan was such that fully three quarters of all the city respondents received and answered messages by shouting through closed doors or by peering out through peepholes. In the towns three quarters of the respondents opened the door.

An experiment to test whether the social anonymity and impersonality of the big city encourages greater vandalism than do small towns was carried out by Zimbardo. He arranged to have left in the street two cars with the license plates removed and the hoods opened as bait for potential vandals. One was left near the Bronx campus of New York University, the other near Stanford University in Palo Alto.

The New York car was stripped of all movable parts within the first 24 hours, and by the end of 3 days was only a hunk of metal rubble. Unexpectedly, however, most of the destruction occurred during daylight hours, usually under the scrutiny of observers, and the leaders in the vandalism were well-dressed white adults. The Palo Alto car was left untouched.[75]

These studies are preliminary attempts to tackle the very thorny problem of what factors constitute the differences between the psychological atmospheres of cities and towns. The use of standard naturalistic experimentation in different areas varying along the dimension of size is a promising start. But even more important is a unifying theoretical concept that helps provide such an orderly approach. The concept of "overload" is suggested by Milgram. This concept may help to explain contrasts in city and town behavior ranging from competition for scarce facilities, such as the subway rush, to the evolution of distinctive urban norms, such as the acceptance of noninvolvement, impersonality, and aloofness in urban life. Milgram suggests that personalities of rural and urban dwellers are not essentially different. Rather, the contrast between rural and urban behavior is basically the result of responses of similar people to entirely different environmental situations. Further studies are needed to test this proposition.

Variations in urban atmosphere

Not only do cities differ from towns but the urban atmosphere of one city may differ considerably from the atmosphere in another. This has been noticed and commented upon by many observers. Martin Meyerson suggests that particular urban forms may be associated with national character.[76] Some examples he discusses are the urban square in Italy, the residential square in England, the boulevard in France, the skyscraper in America, and the neglect of civic design in Japan. In reference to the large, partially enclosed, treeless urban squares of Italy, where large numbers of people regularly congregate, Meyerson comments:

> I suggest that there is in Italy a parallelism between the urban square and a particular kind of gregariousness. In a sense this is only saying something beyond dispute: the Italians who congregate in urban squares have a tendency to congregate in urban squares. The square and the tendency reinforce each other. But without the square, the particular kind of gregarious activity that takes place might not exist; conversely without this trait, the Italians would make less use of the square than they now do.[77]

Over time in this case there has been a gradual development of a design that fits and reinforces the local behavior pattern. A general aim of urban design might be to enhance the possibilities of reinforcing the unique qualities of social behavior that characterize each place. These patterns of social behavior vary considerably. This is indicated by Meyerson who notes that when the Italian urban square was introduced to Britain it was modified markedly to meet English desires. The squares that were developed were no longer Italian in character. They had become quiet London residential squares, with informal, nature-copying gardens surrounded by townhouses. This relationship between landscapes and landscape tastes is explored further in Chapter 7.

Roy Feldman developed a novel approach to measure cross-cultural differences in urban atmospheres.[78] He used a series of naturalistic experiments in comparing the treatment of strangers by Parisians, Athenians, and Bostonians. In each city he compared the treatment of foreigners with that of compatriots. Simple everyday situations provided the means for testing the differences, with a native and a foreign couple serving as experimenters in each city. In a series of separate experiments the investigators asked for directions, asked people to mail a letter for them, provided people with the opportunity of falsely claiming money, overpaid in shops, and compared taxicab charges for foreigners and compatriots. In the taxicab experiment the compatriot and the foreigner would hail cabs in the same city at the same time of day, often within minutes of each other. The compatriot would address the driver in the native language while the foreigner would hand the driver a slip of paper with the destination written

on. The destination was the same, so that the cost of the ride for the foreigner and the compatriot could be directly compared.

The results showed that where differences in treatment occurred, it was the compatriots who were treated better in Boston and Paris. But foreigners were treated better in Athens, where foreign tourists are considered members of the in-group. More important than the results was the example of experimental procedures to measure aspects of everyday behavior. Such experiments compare directly with those used by Sommer that were discussed in Chapter 2. They provide more pertinent measures of urban behavior than are likely to be forthcoming from laboratory experiments.

Another example of a methodology that can be used to measure direct responses to different cities is that developed by David Lowenthal and Marquita Riel.[79] They had several sets of subjects walk along several one-half-mile set paths in New York, Boston, Cambridge, and Columbus. The paths were selected as quintessentially New York, Boston, etc., in flavor. The subjects were asked to record their general impressions and to make judgments about each milieu in terms of 25 attribute pairs previously selected as significant for comparing environmental responses (Table 5.5). Each of the cities was perceived to have a unique set of attributes. New York was described as high-class, fashionable, lively, exciting, entertaining, full of tourists, vulgar, foreign, and dangerous. Boston was seen as old and quaint and "different." Columbus and Cambridge were considered green, parklike, neat, and tidy but also run-down and messy.

When the associations of attributes of the subjects actually experiencing the environment through the walks were compared with those of a control group who did not, interesting contrasts appeared. The mental structure

Table 5.5 Environmental Attribute-pairs Responded to in Urban Walks

1	Natural	Artificial	15	Quiet	Noisy
2	Contrast	Uniform	16	Vivid	Drab
3	People	Things	17	Self-awareness	Awareness of surroundings
4	Ugly	Beautiful			
5	Appearance	Meaning	18	Pleasant	Unpleasant
6	Smelly	Fresh	19	Business use	Living use
7	Vertical	Horizontal	20	Clean	Dirty
8	Ordered	Chaotic	21	Dense	Empty
9	Moving	Motionless	22	Suburban	Urban
10	Smooth	Rough	23	Individual features	Overall views
11	Poor	Rich			
12	Open	Bounded	24	Like	Dislike
13	Boring	Interesting	25	Dark	Light
14	Old	New			

SOURCE: David Lowenthal and Marquita Riel, "Environmental Assessment: A Case Study of New York City," *Publications in Environmental Perception*. No. 1, American Geographical Society, New York, 1972, p. 5. Reprinted by permission.

based on semantic responses differed from that based on observation and experience. The structuring of perception by language noted in Chapter 1 is clearly seen working, as Lowenthal and Riel conclude that

> while language at times reinforces environmental experience, at other times the two are opposed. The differences help to show how and explain why the way we think we see the world is in many respects, not the way we actually do see it.[80]

For example, semantic responses tended to connect people with the adjectives "clean" and "fresh," while in the environmental context people are associated with "smell" and "dirty." It is as if the semantic associations could not stand up to the test of reality. Such findings should be of great interest to those investigating the development of mental maps. For planners a major implication is that we cannot rely solely on semantic-type tests. There is a need for input from human experience and response in the development of design vocabularies, concepts, and building blocks.

Sounds in the city

So far the discussion of urban atmosphere has focused mainly on visual elements and on the press of other people. Another major element among the many that bombard the senses of the city dweller is sound. When annoying or irritating, it is referred to as noise or noise pollution. Much of the work on noise pollution has been concerned with the physical magnitude of the sound. But it is clear that noise, as a social or psychological problem, is not always directly related to physical magnitude. Samuel Klausner points this out in a thoughtful discussion of the topic and also indicates how little attention has been paid to the social and psychological aspects of sound or noise.[81] He suggests that sound is interpreted as unpleasant when it occurs at the wrong time and the wrong place. If it seems necessary, it is apparently less annoying. Our own noise is less unpleasant to us than that generated by other people. Sounds that intrude from outside, which we cannot control or which have no relationship to our activities, are perceived as bothersome. As a general hypothesis Klausner suggests:

> The rates of social interaction follow a rhythmic pattern. At various times people enter the arena of social activity and at other times withdraw from it. There are small day and night rhythms, the larger weekly rhythms and even larger seasonal rhythms. The pattern of bothersomeness of noise seems to constitute a counterpoint to such life rhythms. The same noise seems more bothersome during the evening than during the day, on weekends than on weekdays, and probably more so during the winter than the summer, allowing for differential access of noise through open windows.[82]

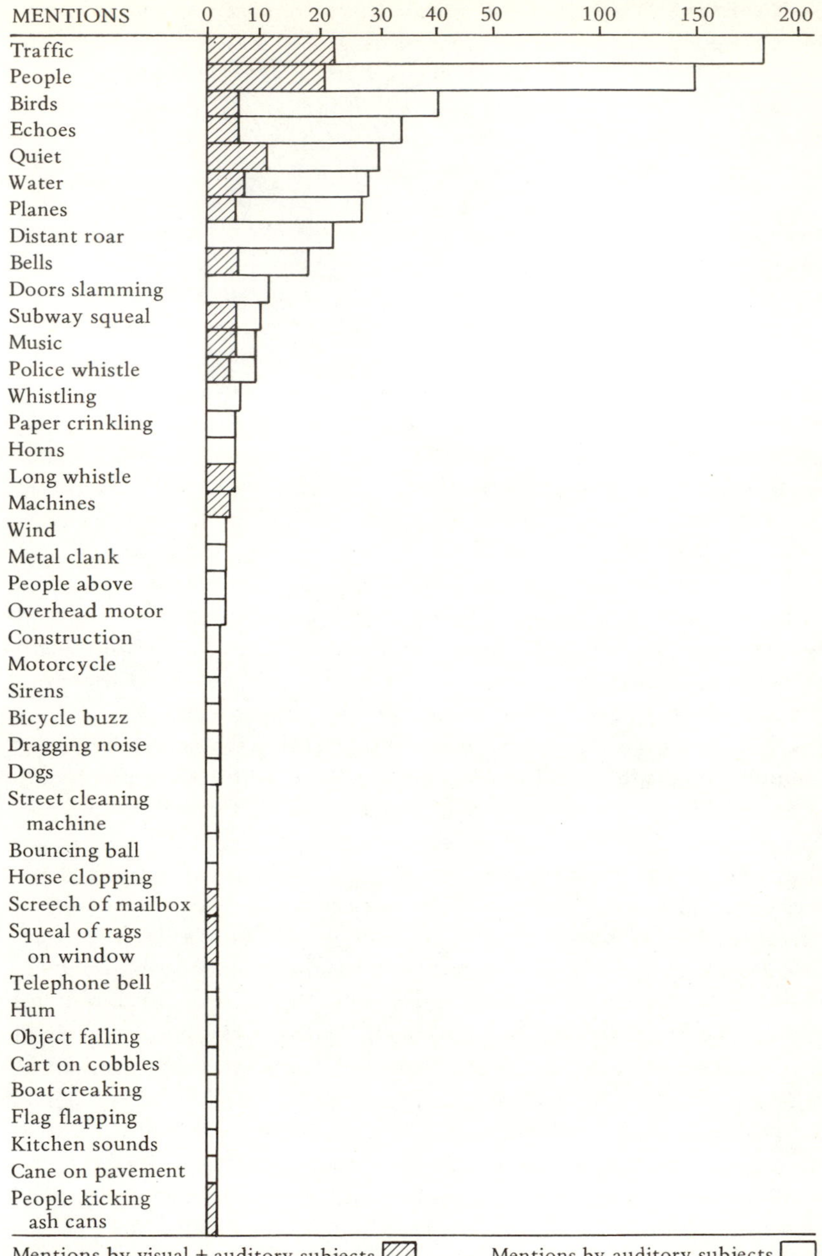

Fig. 5.18 Variety and frequency of mentions of sounds. (From "The Sonic Environment of Cities," by Michael Southworth. Adapted from *Environment and Behavior,* Vol. 1, No. 1 (June, 1969), pp. 49-70 by permission of the publisher, Sage Publications, Inc.)

To alleviate noise pollution as a social and psychological problem, it would not be enough simply to lower the physical level of the sound. In addition, a total analysis of the urban soundscape in terms of temporal rhythms would be necessary to determine which sounds should be screened out, where, when, and for whom. Some ideas as to how this might be done are provided by the research of Michael Southworth.

Southworth, a planner from MIT, carried out an exploratory study of the Boston soundscape.[83] He was not so much concerned with noise pollution as with the perceived variety and character of city sounds and how such sounds influence perception of the visual city. He investigated the uniqueness of local sounds in particular settings and the degree to which the place's activity and spatial form were communicated by sounds. Changes in the soundscape over time and under different weather conditions were also studied.

Southworth's method displayed originality. He took blindfolded subjects on wheelchair trips through the city at different times of the day and week and under varied weather conditions. In an effort to test the interactions between seeing and hearing, he took strange sets of trios on the same trip. Each trio consisted of a strictly auditory subject (blindfolded), a visual subject (who wore earplugs and earmuffs) and a visual-auditory subject with normal hearing and seeing. The trip took about one hour, and throughout its duration the subjects spontaneously described their impressions on portable tape recorders. Later the subjects were asked to draw maps of the trip as they remembered it and to recall and describe their most memorable as well as their most- and least-liked settings.

The results are revealing. Although many different sounds were identified (Figure 5.18), most of the sound settings in the sequence lacked uniqueness and failed to convey an auditory sense of place. But there were five outstanding exceptions, listed in order of dominance as: Washington Street and Filene's corner and their crush of people, sounds, whistles, cars, and Musak (Settings 26, 27, 28), India Wharf and its quiet openness penetrated intermittently by distant planes, bells, ship horns, and sometimes gulls and water (Setting 17), the elevated Central Artery and its constant echoing roar (Settings 14 and 18), Beacon Hill and its array of residential sounds (Settings 1, 33, 34), and the Common and its church bells, people, birds, and open feeling (Settings 29, 30, 31). All these settings are indicated on the map of a portion of the Boston soundscape shown in Figure 5.19.

Southworth provides the planner with many suggestions for improved design. A major starting point would be to reduce and control noise. But, in addition, many steps should be taken to increase the informativeness of the soundscape. Changes are needed to increase the identity of the soundscape, to increase the number of opportunities for delight in sounds, and to increase the correlation among sound, vision, and activity form. Specific places where some of these are needed are noted on the map of the soundscape (Figure 5.19). For example, the waterfront is an area with a

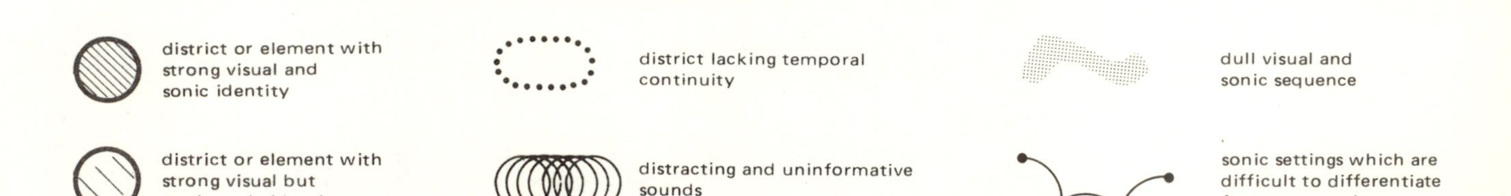

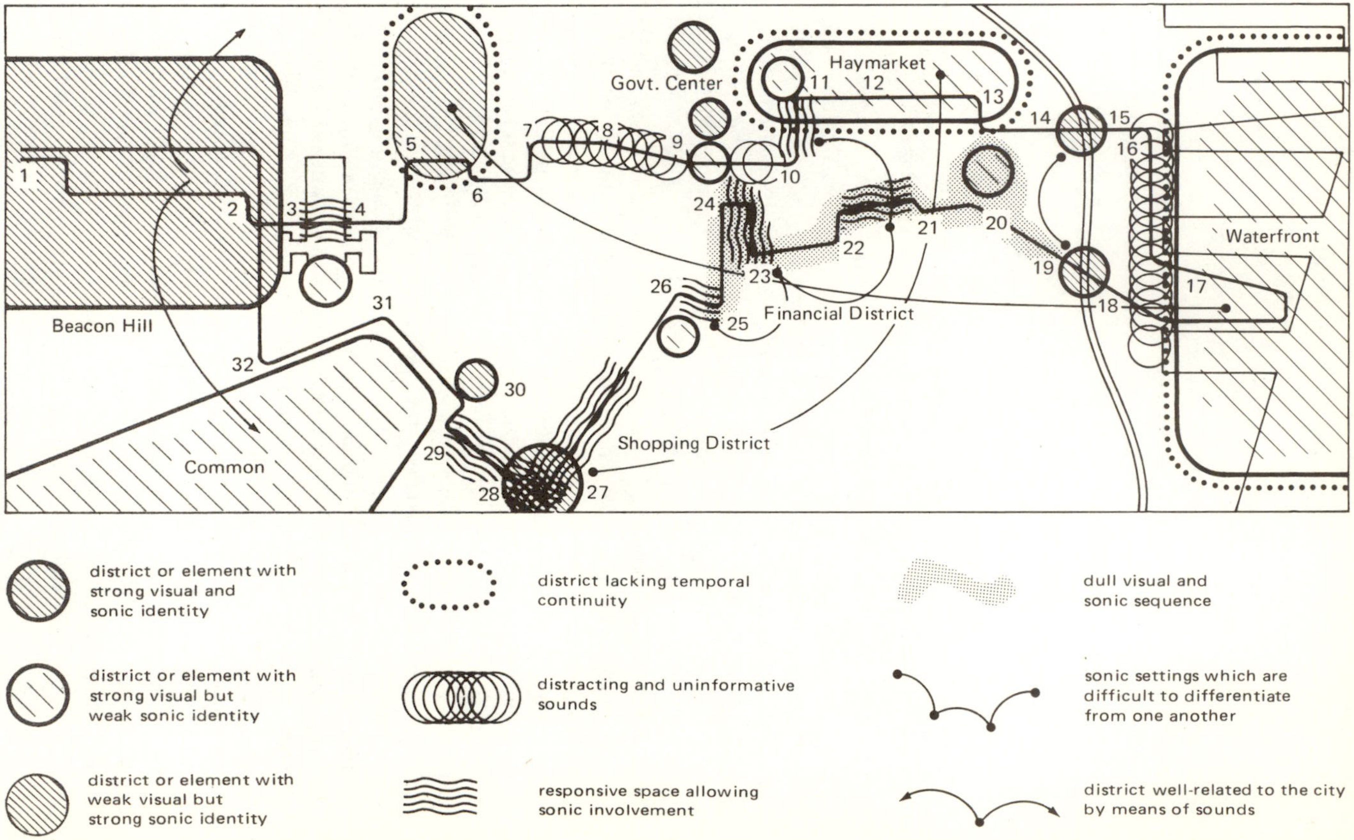

Fig. 5.19 Evaluation of part of Boston soundscape. (From "The Sonic Environment of Cities," by Michael Southworth. Adapted from *Environment and Behavior,* Vol. 1, No. 1, (June, 1969), pp. 49-70 by permission of the publisher, Sage Publications, Inc.)

strong visual but weak sonic identity. Here the sense of place could be enhanced by the addition of new and informative sounds, such as "boats with horns that call out destination-coded sounds."[84] Certain other areas are indicated as responsive spaces because they are usually quiet, sonically responsive, and visually strong. For example, in such places "sequences of sonically differentiated floor materials which squeak, rumble, squish or pop when walked upon would be fun"; another suggestion is for "large animated sculptures which make sound when people move around them"[85] For still other places sonic signs might be used to enliven the scene, provide public information, and strengthen the sense of place.

> Just as chiming clocks tell the time of day, or sirens warn of an emergency, symbolic sounds could be used to inform one of the weather, approaching buses or trolleys, or of special events such as baseball games, concerts or sales. . . . The public sounds of certain districts, such as police whistles or bells, could even be given special character and would strengthen the identity of a locale."[86]

Before leaving the topic of sound in the city, we will turn from Southworth's basically positive approach to sound to examine a negative case of noise. A study by Ronald Cooke and myself provides a link to studies of perception of natural hazards, to be discussed in more detail in the following chapter.[87] Noise pollution was among the problems cited in our study of public perception of environmental quality in Tucson, Arizona. The noise pollution is mainly due to the presence of an Air Force base on the southeastern edge of the city. A questionnaire was completed by a sample of citizens from different parts of the city. The results in relation to noise pollution reveal the parochial manner in which environmental problems may be perceived. Comparison of the locations of the respondents with a map showing noise intensity contours (Figure 5.20) shows that all those in the zone of greatest intensity perceived the noise pollution problem as serious or very serious (Table 5.6). However, the proportions who perceived a noise pollution problem diminish regularly through the zones of lesser intensity. Any action group that wishes to obtain public support to combat noise pollution would be well advised to concentrate their efforts in the high-decibel zones, for elsewhere it is not even considered a problem. A study by the Tucson City Planning Department noted the same parochial pattern of concerns for other urban problems.[88] Air pollution was seen as the major problem by those in the higher areas, who could look down and see it regularly; traffic congestion was the major problem in the eastern zones, where most of the longer-distance commuters live; and urban sprawl was most noticed in suburban ranch areas, where the desert is rapidly disappearing. If planning decisions require broad political support, it is easy to imagine many difficulties in arriving at a consensus on problem priorities. If education to increase public awareness is desired, there are difficult questions regarding the most effective means of carrying out the program.

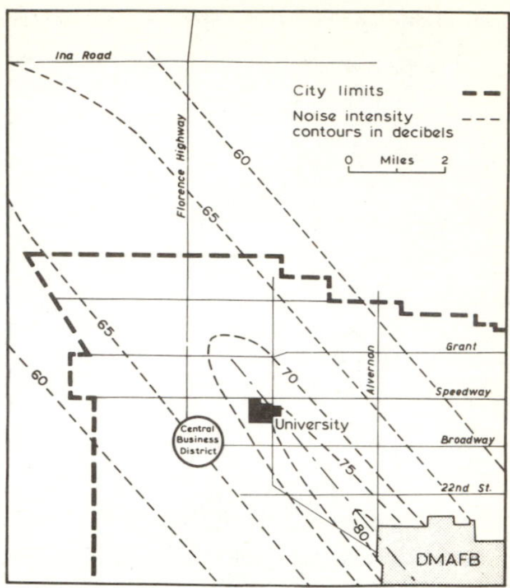

Fig. 5.20 Zones of aircraft noise, Tucson, Arizona. (Thomas F. Saarinen and Ronald Cooke, "Public Perception of Environmental Quality in Tucson, Arizona," *Journal,* Arizona Academy of Science, 1971, VI, 269. Reprinted by permission.)

Summary

Research on perception of and behavior in the urban environment has strong links to other scales, both smaller and larger. Since the urban area is perhaps the largest-sized segment of the environment that can be seen as a whole and known intimately, research here has many links to investigation of smaller-sized areas. But because of the city's size and complexity, much of the research also resembles that focusing on larger areas. A key aspect of the change that occurs at about the scale of the city is shift from concern with direct perception of the environment to focus on people's conceptions of the environment.

The possibility of measuring people's direct perception of environment is still present at the scale of the city but it becomes less practical at broader scales. As we saw in this chapter, the studies of Southworth and of Lowenthal and Riel were based on taking people through a portion of the city and recording their direct impressions of the experience. Southworth was concerned with the sounds of the city, Lowenthal and Riel mainly with the visual factor. A provocative aspect of the latter studies was the comparison of the semantic linkages of the groups who directly perceived the city with those of others who did not. This comparison suggested the possibility of separating what is seen into the contributions of direct experience and those of previous impressions derived from linguistic factors.

Table 5.6 Rating of Seriousness of Noise Pollution According to Noise Level

Noise level in decibels	Very serious (Number)	(%)	Serious (Number)	(%)	Not serious (Number)	(%)	No answer (Number)	(%)
Over 80	3	50	3	50	0	0	0	0
75-79	0	0	0	0	0	0	0	0
70-74	11	55	7	35	2	10	0	0
65-69	10	39	24	54	2	7	0	0
60-64	11	18	20	32	30	48	1	2
Less than 60	3	12	5	19	18	69	0	0
0	2	7	3	11	20	74	2	7

SOURCE: Thomas Saarinen and Ronald U. Cooke, "Public Perception of Environmental Quality in Tucson, Arizona" *Journal of Arizona, Academy of Science*, 1971, VI, 265. Reprinted by permission.

The staging of naturalistic experiments as a means of measuring variations between the behavior in towns and cities was described by Milgram, while Feldman used similar experiments to try to measure the differences between cities. Such experiments bear a strong resemblance to those staged by Sommer at the scale of personal space. By entering the real-life situation in a normal manner, the experimenter is able to observe directly the behavior stimulated. These simple, often ingenious, and effective means of measuring behavior were discussed in some detail in Chapter 2. Although the unit of environment has changed, the psychological approach is still recognizable in the focus on individual reactions to the environment consisting mainly of other people.

In linking types of behavior with physical locations, the work on urban activity patterns corresponds directly with other studies of activities in smaller areas. What is being sought is a unit of behavior that can be observed, recorded, and compared. Such a unit could serve both the social scientist's interest in behavior and the designer's need for data translatable to form. Constance Perin suggests the concept of *behavior circuits*.

> The study of the man-machine relationship, when fingers, eyes, and elbows function with small-scale equipment, is called ergonomics; what behavior circuit implies is an anthropological ergonomics, tracking people's behaviors through the fulfillment of their everyday purposes at the scale of the room, the house, the block, the neighborhood, the city in order to learn what resources—physical and human—are needed to support, facilitate, or enable them.[89]

A major advantage of a concept such as "behavior circuits" is that it provides a means of assessing the sources of stress or strain in the daily round of activities. Physical changes can then be designed to overcome or minimize the difficulties or, positively, to enhance the desired behavior. Meyerson's description of gregarious congregating in the Italian town square showed at the city level how a particular type of behavior can be reinforced by an appropriate design. The same positive reinforcing or

enabling quality could be designed into environments serving other functions and at many other scales. An intriguing idea is that of replication of successful designs at several different scales, as was described by Wood. In his study of San Cristobal, he found that the replication of form aided in the transfer of skills from the lower to the higher levels in the city's hierarchy of places.

At the scales of the previous chapters, the environment of interest was mainly that consisting of other people or of the built environment of rooms, buildings, or site plans. These interests continued at the scale of the city. In Milgram's concept of "overload," much of the excessive stimulation experienced by urban dwellers is due to the pressure of other people. The built environment was also a major focus of attention, as in Appleyard's inquiry as to why buildings are known. However, in addition to the physical dimensions of the city, greater attention was devoted to symbolic aspects of the urban environment. Even at the level of the room, it was noted that chairs may acquire symbolic meanings. Other examples could be provided at the scale of architectural space or neighborhoods as well. But as one proceeds to very large segments of the environment, the importance of symbolic aspects increases. Images, attitudes, ideas, and mental maps tend to replace direct perception of places and observation of behavior as the focus of research.

The research utilizing sketch maps of various areas may be used to illustrate the transition from awareness of reality in small areas to symbolic representations of very large areas. Sketch maps of environments of various scales can be drawn easily by most people in our society. But, at the scale of the neighborhood or campus, most people will have directly seen the major features in the environment and can perhaps mentally picture them as they sketch their maps. Much of their knowledge is based on direct, often repeated experience in the environment. This is still possible, though not quite so easy, at the scale of the city. In sketching a map of the world, direct experience is no longer of much assistance. Instead, the person must rely on other people's information as embodied in maps. Although most atlas maps are based on accurate information, the problems of representing a globe on a flat surface lead to many distortions that may not be understood by any but the cartographically sophisticated. The likelihood of error increases as the possibility of direct experience decreases. To overcome the most likely types of distortions, more information on the general process of mental map formation is required.

Much discussion in this chapter was devoted to the topic of mental maps. An understanding of people's cognitive representations of the environment is important at all scales because they enable us to orient ourselves and provide a basis for our spatial behavior. More material on mental maps will appear in the chapters ahead. The reason for this is that in the larger areas, there are fewer alternative methods of researching how people perceive the environment.

An important problem is the question of how to measure or study mental maps. The most direct way would be simply to ask people to sketch maps of various areas, as in the studies on the image of the city. Other methods have also been used to investigate various dimensions of mental maps. Downs, as well as Lowenthal and Riel, used the semantic differential technique. Lee and researchers at the Ohio State University asked people for estimates of distances. Horton and Reynolds used questionnaires to determine degree of familiarity with parts of the city.

The map-sketching technique is particularly flexible for exploring various aspects of mental maps. In addition, it has the advantage of providing a comprehensive view with all the elements arranged in relation to one another, as in the real world. The subjective map can also be compared directly to objective measures of the same area. As an essentially projective technique, it has great potential for cross-cultural studies. For these reasons, as well as for its ease of application, planners, educators, and students should find the sketch-map technique useful as a starting point in examining how people perceive their environment. Many further extensions of the technique are possible. Not only is it possible to have people sketch maps at any scale, but it seems likely that maps could be produced to illustrate any psychological dimension as well.[90] Preferences for places or ideal areas have already been investigated. It would also be possible to map a city, a country, or the world in terms of such things as perceived level of danger. Different groups would likely have different perspectives. Comparisons of perceived danger zones of blacks and whites, right-wing and left-wing radicals, hardhats and freaks might be revealing. The maps produced provide a starting point, for they raise a host of questions. A basic one is why certain areas appear while others do not. Much of the ongoing research in city images attempts to answer that question.

Another topic of major importance raised in this chapter is the question of the ages and stages at which knowledge of the environment or mental maps are acquired. From the opening example of the Horton and Reynolds study to the closing one of Cooke and myself, many instances were cited of parochial views of the city and its problems held by local residents. Yet we know next to nothing about how these views develop. Are there optimal ages and stages during which people's views of the city, country, or world could be enlarged or extended? If so, what are they? And how can this be accomplished? Such questions remain foremost in the chapters that follow.

Knowledge of how people perceive and behave in the urban environment becomes more meager as one moves to the largest cities. The comments of Simmel, Plant, and Wirth ring true but are difficult to test directly. What is needed is a theoretical approach that can provide a consistent research direction, such as the concept of "overload" suggested by Milgram. Naturalistic experiments could be designed to measure behavioral changes along defined dimensions, such as size or population density. Totally controlled experiments are difficult in the metropolis; but naturalis-

tic experiments could yield useful behavioral data if staged in carefully selected areas that contrast in one dimension but are otherwise similar. This type of approach was seen in the work of Newman discussed in the Chapter 4.

Another direction of investigation of behavior in the city involves detailed studies of particular elements. In this chapter sound or noise was discussed briefly. It was noted that much remains to be learned about the social and psychological aspects of noise and that these are not always directly related to the physical magnitude of sound. Crowding is another element of city life that should be researched more thoroughly from a subjective point of view.

Though the city may be regarded as an independent system, it also has external relations that tie it to the surrounding region, to the nation, and to the world. Simmel spoke of this functional extension as the most significant characteristic of the metropolis. Other aspects of the relationship of the city to the broader world beyond were suggested in Meyerson's ideas on the relationship between national character and urban design, in Feldman's cross-cultural comparison of some psychological dimensions of urban atmosphere, and in Cooke's and my study of the public perception of environmental quality. As we turn to consider broader conceptual regions, the reader may bear in mind that one such region is the city-region, the area in which the city exerts a clearly dominant influence. And in its turn, the nation may be considered a mosaic of such regions.

Notes

1. See, for example, Leo F. Schnore, "The City as a Social Organism," *Urban Affairs Quarterly*, 1, No. 5 (March, 1966), 58-69; also reprinted in Larry S. Bourne, *Internal Structure of the City: Readings on Space and Environment*, Oxford University Press, New York, 1971, pp. 32-39.
2. For discussion of this, see Bourne, *Internal Structure of the City.*
3. This is the theme of a conference commemorating the fiftieth anniversary of The American Institute of Planners. The results are reported in William R. Ewald, Jr., ed., *Environment for Man: The Next Fifty Years*, Indiana University Press, Bloomington, Ind., 1967.
4. C. A. Doxiadis, "A City for Human Development," *Ekistics*, 25, No. 151 (June, 1968), 380. This entire issue is devoted to anthropics: people in relation to their settlements.
5. Paul and Percival Goodman, *Communitas: Means of Livelihood and Ways of Life*, Vintage Books, New York, 1947.
6. Christopher Alexander, "The City as a Mechanism for Sustaining Human Contact," in Ewald, ed., *op. cit.*, pp. 60-102.
7. *Ibid.*, p. 68.
8. Frank E. Horton and David R. Reynolds, "Action Space Differentials in Cities," in Harold McConnell and David Yaseen, eds., *Perspectives in Geography: Models of Spatial Interaction*, Northern Illinois University Press, Dekalb, Ill., 1971, pp. 83-102, quote on p. 86.

9. F. Stuart Chapin, Jr., and Richard K. Brail, "Human Activity Systems in the Metropolitan United States," *Environment and Behavior,* 1, No. 2 (December, 1969), 107-130, quote on p. 108. See also F. Stuart Chapin, Jr., and Henry C. Hightower, "Household Activity Patterns and Land Use," *Journal of the American Institute of Planners,* 31, No. 3 (August, 1965), 222-231.
10. J. Rannels, *The Core of the City,* Columbia University Press, New York, 1956.
11. J. Douglas Porteous, "The Burnside Teenage Gang: Territoriality, Social Space, and Community Planning," in *Residential and Neighborhood Studies in Victoria,* C. N. Forward, ed., Department of Geography, University of Victoria Western Geographical Series No. 5, University of Victoria, Victoria, B.C., 1973, pp. 130-148.
12. Rannels, *op. cit.,* p. 11.
13. *Ibid.,* p. 19.
14. Chapin and Brail, *op. cit.*
15. Paul-Henry Chombart De Lauwe, *Des Hommes et Des Villes,* Payot, Paris, 1965; see Chapter 2.
16. Jane Jacobs, *The Death and Life of Great American Cities,* Vintage Books, New York, 1961; see, for example, pp. 59-62. See also excerpts in Proshansky *et al., Environmental Psychology,* pp. 312-319, 382-386.
17. Anselm Strauss, "Life Styles and Urban Space," in Proshansky *et al., Environmental Psychology,* pp. 303-12, quote on p. 307; originally published in *Images of the American City,* Free Press, New York, 1961.
18. Willem F. Heinemeyer, "The Urban Core as a Centre of Attraction," in *Urban Core and Inner City,* E. J. Brill, ed., Leiden, Netherlands, 1967, pp. 82-99.
19. For the best single source on the American city, see Strauss, *The American City: A Sourcebook of Urban Imagery,* Aldine Publishing Co., Chicago, 1968.
20. Morton White and Lucia White, *The Intellectual Versus the City,* Mentor Books, New York, 1962.
21. Morton White, "Two Stages in the Critique of the American City," in Oscar Handlin and John Burchard, eds., *The Historian and the City,* M.I.T. Press, Cambridge, Mass., 1963, pp. 84-94, quote on p. 86.
22. See the discussion on this point in Eugene V. Rostow and Edna G. Rostow, "Law, City Planning and Social Action," in Leonard J. Duhl, ed., *The Urban Condition,* Clarion Books, New York, 1963, pp. 357-373.
23. Carl E. Schorske, "The Idea of the City in European Thought," in Handlin and Burchard, eds., *op. cit.,* pp. 95-114.
24. Strauss, "Urban Perspectives: New York City," in *The American City,* p. 5.
25. Kevin Lynch, *The Image of the City,* M.I.T. Press, Cambridge, Mass., 1960.
26. An immediate predecessor was Kenneth E. Boulding, *The Image,* University of Michigan, Ann Arbor, Mich., 1956.
27. See, for example, the sequence of development of the city as viewed from roads and highways, starting with Lynch then D. Appleyard, K. Lynch, and J. R. Meyer, *The View from the Road,* M.I.T. Press, Cambridge, Mass., 1964, and most recently Stephen Carr and Dale Schissler, "The City as a Trip: Perceptual Selection and Memory in the View from the Road," *Environment and Behavior,* 1, No. 1 (June, 1969), 7-35.
28. See, for example, Philip Thiel, "Notes on the Description, Scaling, Notation, and Scoring of Some Perceptual and Cognitive Attributes of the Physical Environment," in Proshansky *et al., Environmental Psychology,* pp. 593-619, and Lawrence Halprin, "Motation," *Progressive Architecture,* 46 (July, 1965), 126-133.
29. Derk DeJonge, "Images of Urban Areas: Their Structure and Psychological Foundations," *Journal of the American Institute of Planners,* 28 (November, 1962), 266-276.

30. John Gulick, "Images of an Arab City," *Journal of the American Institute of Planners,* 29, No. 3 (1963), 179-198.
31. Thomas F. Saarinen, *Image of the Chicago Loop,* 1964, and *Image of the University of Arizona Campus,* 1967.
32. Hans-Joachim Klein, "The Delimitation of the Town-Centre in the Image of its Citizens," in University of Amsterdam Sociographical Department, E. J. Brill, ed., *Urban Core and Inner City,* Leiden, Netherlands, 1967, pp. 286-306.
33. Mahmoud Zawawi, "Perception of Downtown: A Case Study of Washington, D.C." (Master's thesis, George Washington University, 1970).
34. Heinemeyer, *op. cit.;* see Footnote 18.
35. Robert W. Kates, "Human Perception of the Environment," *International Social Science Journal,* 22, No. 4 (1970), 648-660.
36. Because so many of the studies appear as unpublished theses and dissertations or as publications of local planning departments, or in foreign countries, it is virtually impossible to keep up with them. See, for example, Footnote 33 above, or Department of Community Development, *Tucson Visual Environment,* Department of Community Development, Tucson, 1969. For a listing of many British examples, see Brian Goodey, Alan W. Duffett, John R. Gold, and David Spencer, *City Scene: An Exploration into the Image of Central Birmingham as Seen by Area Residents,* Research Memorandum No. 10, The University of Birmingham Centre for Urban and Regional Studies, October, 1971.
37. Denis Wood, *Fleeting Glimpses,* Clark University Cartographic Laboratory, Worcester, Mass., 1971.
38. Donald Appleyard, "City Designers and the Pluralistic City," in Lloyd Rodwin and Associates, *Planning Urban Growth and Regional Development: The Experience of the Guayana Program of Venezuela,* M.I.T. Press, Cambridge, Mass., 1969, pp. 422-452.
39. Appleyard, "Why Buildings Are Known: A Predictive Tool for Architects and Planners," *Environment and Behavior,* 1, No. 2 (December, 1969), 131-156.
40. Appleyard, "Styles and Methods of Structuring a City," *Environment and Behavior,* 2, No. 1 (June, 1970), 100-116.
41. Appleyard, "Why Buildings Are Known," 154.
42. *Ibid.,* 155.
43. Appleyard, "Notes on Urban Perception and Knowledge," in John Archea and Charles Eastman, *EDRA Two,* Proceedings of the Second Annual Environmental Design Research Association Conference, Pittsburgh, Pa., October, 1970, p. 99.
44. Appleyard, "Styles and Methods of Structuring a City," 111.
45. *Ibid.,* 113.
46. See the following collections of papers: David Stea and Roger Downs, "Cognitive Representations of Man's Spatial Environment," entire issue of *Environment and Behavior,* 2, No. 1 (June, 1970); Downs and Stea, Chairmen, Session Two, "Environmental Cognition and Behavior," in Archea and Eastman, *EDRA Two,* pp. 95-142; Downs and Stea, *Image and Environment: Cognitive Mapping and Spatial Behavior,* Aldine Publishing Co., Chicago, 1973.
47. Stea and Downs, "From the Outside Looking in at the Inside Looking Out," *Environment and Behavior,* 2, No. 1 (June, 1970), 3-12.
48. See Footnote 73, Chapter 4.
49. Downs, "The Cognitive Structure of an Urban Shopping Center," *Environment and Behavior,* 2, No. 1 (June, 1970), 13-39.
50. *Ibid.,* 38.
51. Terence Lee, "Perceived Distance as a Function of Direction in the City," *Environment and Behavior,* 2, No. 1 (June, 1970), 40-51.
52. R. G. Golledge, R. Briggs, and D. Demko, "The Configuration of Distances in

Intra-Urban Space," *Proceedings of the Association of American Geographers,* 1 (1969), 60-65.

53. R. G. Golledge, R. Briggs, and D. Demko, "Cognitive Approaches to the Analysis of Human Spatial Behavior," in William H. Ittelson, ed., *Environment and Cognition,* Seminar Press, New York, 1973, pp. 59-94.
54. See, for example, Thomas Sieverts, "Perceptual Images of the City of Berlin," in Brill, ed., *Urban Core and Inner City,* pp. 282-285; Florence C. Ladd, "Black Youths View Their Environment: Neighborhoods Maps," *Environment and Behavior,* 2, No. 1 (June, 1970), 74-99; or the comments of Charles Tilly, "Anthropology on the Town," *Habitat,* 10, No. 1 (January-February, 1967), 20-25. See also Alvin K. Lukashok and Kevin Lynch, "Some Childhood Memories of the City," Journal of the American Institute of Planners (Summer, 1956), 142-152; and Robert Maurer and James C. Baxter, "Images of the Neighborhood and City Among Black, Anglo-, and Mexican-American Children," *Environment and Behavior,* 4, No. 4 (December, 1972), 351-388.
55. Roger A. Hart and Gary T. Moore, *The Development of Spatial Cognition: A Review,* Place Perception Research Report No. 7, Graduate School of Geography and Department of Psychology, Clark University, Worcester, Mass., 1971.
56. Some of the earlier reports are J. M. Blaut, *Studies in Developmental Geography,* Place Perception Research Report No. 1, Graduate School of Geography, Clark University, Worcester, Mass., October, 1969; Merrie Ellen Muir, *The Use of Aerial Photographs as an Aid in Teaching Map Skills in the First Grade,* Place Perception Research Report No. 3, Graduate School of Geography, Clark University, Worcester, Mass., July, 1970; J. M. Blaut and David Stea, *Place Learning,* Place Perception Research Report No. 4, Graduate School of Geography, Clark University, Worcester, Mass., December, 1969; Blaut and Stea, *Notes Toward a Theory of Environmental Learning,* Place Perception Report No. 8, Graduate School of Geography, Clark University, Worcester, Mass., March, 1970.
57. James M. Blaut, F. McCleary, Jr., and America S. Blaut, "Environmental Mapping in Young Children," *Environment and Behavior,* 2, No. 3 (December, 1970), 335-349.
58. *Ibid.,* p. 340.
59. *Ibid.,* p. 347.
60. See, for example, the discussion in Doxiadis, *op. cit.,* pp. 374-394.
61. William Michelson, *Man and His Urban Environment: A Sociological Approach,* Addison-Wesley, Reading, Mass., 1970, Chapter 7.
62. James S. Plant, "The Personality and an Urban Area," in Paul K. Hatt and Albert J. Reiss, eds., *Cities and Society,* Glencoe Press, New York, 1957, pp. 647-665.
63. Georg Simmel, "The Metropolis and Mental Life," in Hatt and Reiss, *op. cit.,* pp. 635-646; quote on p. 639.
64. Louis Wirth, "Urbanism as a Way of Life," in Albert J. Reiss, ed., *Louis Wirth: On Cities and Social Life,* University of Chicago Press, Chicago, 1964, pp. 60-83.
65. *Ibid.,* p. 71.
66. Simmel, *op. cit.,* p. 635.
67. *Ibid.,* p. 642.
68. *Ibid.,* p. 642.
69. Jacobs, *op. cit.*
70. Herbert J. Gans, "Urban Vitality and the Fallacy of Physical Determinism," in *People and Plans: Essays on Urban Problems and Solutions,* Basic Books, New York, 1968, pp. 25-33.
71. Lewis Mumford, "Home Remedies for Urban Cancer," in *The Urban Prospect,* Harcourt Brace Jovanovich Inc., New York, 1968, pp. 182-207; quote on p. 191.
72. *Ibid.,* p. 207.

73. Stanley Milgram, "The Experience of Living in Cities," *Science*, 167, No. 13 (March, 1970), 1461-1468; quote on p. 1462.
74. *Ibid.*, p. 1462.
75. *Ibid.*, p. 1464.
76. Martin Meyerson, "National Character and Urban Development," in C. J. Friedrich and S. E. Harris, eds., *Public Policy*, Vol. 12, Harvard Graduate School of Public Administration, Cambridge, Mass., 1963, pp. 78-96.
77. *Ibid.*, p. 87.
78. Roy E. Feldman, "Reponse to Compatriot and Foreigner Who Seek Assistance," *Journal of Personality and Social Psychology*, 10, No. 3 (1968), 202-214.
79. David Lowenthal and Marquita Riel, *Publications in Environmental Perception*, American Geographical Society, New York, 1972; eight separate reports.
80. Lowenthal and Riel, "The Nature of Perceived and Imagined Environments," *Environment and Behavior*, 4, No. 2 (June, 1972), 198-207; quote on 206.
81. Samuel Z. Klausner, *On Man in His Environment*, Jossey-Bass, San Francisco, 1971, Chapter 6.
82. *Ibid.*, p. 122.
83. Michael Southworth, "The Sonic Environment of Cities," *Environment and Behavior*, 1, No. 1 (June, 1969), 49-70.
84. *Ibid.*, p. 68.
85. *Ibid.*, p. 68.
86. *Ibid.*, p. 69.
87. Thomas F. Saarinen and Ronald U. Cooke, "Public Perception of Environmental Quality in Tucson, Arizona," *Journal of the Arizona Academy of Science*, 6 (1971), 260-274.
88. City of Tucson Planning Division, *Community Issues*, Comprehensive Planning Report No. 6 (March, 1973).
89. Constance Perin, *With Man in Mind: An Interdisciplinary Prospectus for Environmental Design*, M.I.T. Press, Cambridge, Mass., 1970, p. 77.
90. Milgram, "Introduction to Chapter 2," in William H. Ittelson, *Environment and Cognition*, Seminar Press, New York, 1973, pp. 21-27.

Six

Large conceptual regions

"At the heart of managing a natural resource is the manager's perception of the resource and of the choices open to him in dealing with it."
Gilbert F. White

The scale of studies in this chapter is that of climatic regions; physical regions, such as river basins; political regions, such as provinces or state parks; and cultural or agricultural regions. The size may vary greatly but tends to be less large and complex than a country but broader in area than a city. The numbers and types of such regions are limited only by our own imagination in defining them. Perception of environment at this scale and at the scale of the even larger areas that follow could really be more accurately termed "conceptions of environment." The slice of the environment considered is generally too large to be perceived by the senses all at once, and thus what are usually studied are the mental images, ideas, or attitudes people have about selected aspects of the region under investigation. If one judges by the number of studies that have been completed, this is the scale most congenial to geographers interested in environmental perception and behavior, and most of the material in this chapter reviews their research. A change that occurs is a shift in emphasis from the built and social environments to perceptions and behavior in relation to the natural environment.

Roots of perception studies in geography

Perception of environment is part of the larger system of human beings and environment that has always been a major concern of geography. Therefore, it is not surprising that the roots of the present interest in perception of environment run deep in this discipline. These roots appear in such diverse themes as climatic influences, cultural appraisal, regional consciousness, and regional description. The ideas of climatic influences and environmental determinism derive from the perceived power of the natural environment in conditioning our use of the earth and even our forms of society. Such ideas appear in the oldest written works. *Cultural appraisal* is a term long used by geographers and anthropologists to describe the differing perceptions of various culture groups in their assessment of the resources of their regions.[1] These ideas are considered later in this chapter. *Regional consciousness* is an awareness of the uniqueness of an area involving a feeling of belonging or sense of shared existence found among inhabitants of the region.[2] It is discussed in more detail in Chapter 7.

Geographical description also involves environmental perception, as what is included or emphasized depends upon individual selection.[3] Later in this chapter we will consider some studies that focus on accounts of explorers, early settlers, and others to gain insight into impressions of particular regions in past periods. Analysis of such material may reveal the degree to which these impressions are a product of preconceived notions and also whether such ideas influenced the subsequent settlement and development of the regions. Some of the most successful examples of regional description have appeared in regional novels. Such writers as Joseph Conrad, Thomas Hardy, and Ole Edvart Rölvaag conjure up a vivid sense of place and express in a profound manner some of the psychological dimensions involved in the people-environment interaction. The utility of novels in assessing American attitudes toward small towns and cities has already been noted.

The value of novelists' insights has long been recognized by geographers, who have at times authored articles on particular novelists and their regions.[4] E. W. Gilbert has mapped almost 150 such novelists' territories in England and Wales,[5] and Otis Coan and Richard Lillard's book contains an annotated list of regional novels of the United States, Mexico, and Canada.[6] But this is only one kind of geographical description. Others include travelers' accounts, serious essays that aim to analyze the character of regions,[7] textbooks, and landscape paintings.[8] All reveal aspects of past and present perceptions of the environment, and all will be noted as sources for studies of this type.

Because kindred notions have always been a part of the geographical tradition, it is difficult to identify a single starting point for the current surge of interest in environmental perception. However, this diversity of

roots does make it easy for geographers to accept perception studies, for, at least when viewed from the present perspective, many of their most distinguished predecessors seemed to have advocated exactly this type of approach. Such pleas can be seen in quotes culled from the works of previous generations of geographers in special fields ranging from natural resources to historical geography. But more appeared than an odd quote or two.

In addition to being credited with the first plea in American geography for the study of geographical perception, John Kirtland Wright in 1925 produced a monograph showing exactly how such a concept could be applied.[9] This book, *The Geographical Lore of the Time of the Crusades,* is considered further in the discussion of world views in Chapter 8. Unfortunately, the concept and method were ignored until 1943, when Ralph Brown's *Mirror for Americans* was published.[10] In 1946 Wright tried again, this time in his presidential address to the Association of American Geographers, where he introduced the term "geosophy," which

> . . . covers the geographical ideas, both true and false, of all manner of people—not only geographers, but farmers and fishermen, business executives and poets, novelists and painters, Bedouins and Hottentots—and for this reason it necessarily has to do in large degree with subjective conceptions.[11]

Geosophy did not catch on, nor did the term "behavioral environment," proposed by William Kirk, whose statement, framed in terms of gestalt psychology, was the first in geography to separate the perceived environment as a distinct surface.[12] But as H. C. Brookfield observed,[13] it has really only achieved retrospective notice, and the real bridge between the past and present approaches to perception in geography was the subject of a review paper by David Lowenthal that appeared in 1961.[14] Since about that time there has been a steadily expanding flow of publications on this theme.

Perception of natural hazards

By far the most numerous and well-established set of perception studies in geography are those derived from the interests of geographers at the University of Chicago in perception as a factor in resource management. A constant theme in this research has been the concern for practical applications. Indeed, perception studies were developed in response to the need for more thoughtful solutions to the problems of dealing with natural hazards. Gilbert F. White began by investigating the factors affecting human adjustment to flood plains.[15] Companion volumes by Robert William Kates and White contain a good review—and much of the flavor—of the flood plain studies.[16] Work on floods led to questions about other types of natural hazards. The entire sequence of development can be sum-

marized as flood plain studies; natural hazards in general[17]; detailed studies of specific hazards such as coastal storms,[18] drought,[19] snow,[20] landslides,[21] tidal waves,[22] frost,[23] tropical cyclones,[24] and earthquakes[25]; attitudes toward environment[26]; attitudes toward water[27] and the atmosphere[28]; the role of perception in specific water uses[29]; perception of artificial hazards, such as air and water pollution[30]; and, most recently, examination of several hazards at a place[31] and comparison of adjustments to specific hazards over a wide range of world sites.[32]

The magnitude of the problem of natural hazards may be gauged by the loss of life and the damages they cause. Table 6.1 indicates by disaster type the global loss of life due to natural hazards during the twenty-one-year period 1947-1967. The data were compiled for all major world disasters, defined as causing at least one million dollars' damage or at least 100 people killed or injured. Table 6.2 shows the injuries, loss of life, and damages on an annual basis for the United States.

The geographic work on perception of natural hazards has been ably summarized by the principal researchers, Ian Burton, Robert Kates, and Gilbert White.[33] They describe the paradox presented by growing damages due to natural hazards, even as we become more able to manipulate or control certain aspects of nature. Geographers have long been curious about the behavior of people who persistently return to resettle areas after devastation by floods, volcanoes, earthquakes, droughts, or other natural

Table 6.1 Loss of Life by Disaster Type, 1947-1967

Disaster types	Number of lives lost	Percent of total loss of life
Floods	173,170	39.2
Rain	1,100	0.2
Gale and thunderstorms	20,940	4.7
Blizzards and snowstorms	3,520	0.8
Sand, dust storms	10	-
Cyclone and tidal waves	89,440	20.2
Tidal waves	3,180	0.7
Hurricanes	13,225	3.0
Tornado groups	3,395	0.8
Typhoons	52,400	11.9
Hailstorms	-	-
Heat-waves	4,675	1.1
Cold-waves	3,370	0.8
Fog	3,550	0.8
Earthquakes	56,100	12.7
Volcanoes	7,220	1.6
Avalanches	3,680	0.8
Landslides	2,880	0.7
TOTAL	441,855	100.0

SOURCE: Lesley Sheehan and Kenneth Hewitt, "A Pilot Survey of Global Natural Disasters of the Past Twenty Years," *Natural Hazard Research Working Paper No. 11*, Department of Geography, University of Toronto, 1969, p. 18. Reprinted by permission.

Table 6.2 Injuries, Loss of Life and Property Damage Resulting From Selected Natural Hazards in the United States, Annual Basis: 3- to 5-Year Average

Hazard	Injuries	Loss of life	Property damage (millions of $)
Hurricanes	6,755	41	448.7
Tornadoes	2,019	124	180.0
Excessive heat		236	
Winter storms (excessive cold)	500	366	182.1
Lightning	248	141	33.5
Floods	610	62	399.5
Earthquakes	112	28	102.7
Tsunamis	40	24	21.0
Transportation accidents related to weather, etc.	237	288	18.9
Drought			78.6*
Hail			22.1*
Excess Moisture			27.7*
Wind			11.8*

*Farm crop losses only. SOURCE: Chart prepared by the United States Dept. of Commerce, Environmental Science Services Administration (1968). From Ian Burton and Kenneth Hewitt, *Hazardousness of a Place*, Department of Geography, University of Toronto, 1971, p. 8. Reprinted by permission.

disasters. Although it seems likely that modern peoples are more aware of the risks of repeated disasters, the reinvasion of hazard zones probably continues as in the past. In fact, probably even more pressure is placed on these areas as population numbers grow and human beings and their works continue to spread over the earth. And, with increasing pressure on resources, a more delicate adjustment to nature becomes necessary.

The approach is ecological, and the inhabitants' view is sought in an effort to understand the long-range human adjustment to the hazard. This provides the possibility of assessing our ability to perceive and understand the world around us and to choose appropriate courses of action. Improved public policies may be achieved by offering education and information in areas where knowledge is deficient. In many cases this could mean providing people with a broader view of the complete range of theoretically possible adjustments. As seen in Table 6.3, there are many more possibilities than the prevailing public approach of offering immediate relief followed by a technological solution, as when dams follow floods and irrigation projects follow droughts. Clearly, there are many alternatives. The problem of flood damage, for example, might be more wisely handled by preventing dense development on flood plains through flood-plain zoning rather than by relying on the commonly perceived technological solutions of bigger dams and levees. Reliance on technological solutions

Table 6.3 Theoretical Range of Adjustments to Geophysical Events

Class of adjustment	Events		
	earthquakes	**floods**	**snow**
Affect the cause	No known way of altering the earthquake mechanism	Reduce flood flows by: land-use treatment; cloud seeding	Change geographical distribution by cloud seeding
Modify the hazard	Stable site selection: soil and slope stabilization; sea wave barriers; fire protection	Control flood flows by: reservoir storage; levees; channel improvement; flood fighting	Reduce impact by snow fences; snow removal; salting and sanding of highways
Modify loss potential	Warning systems; emergency evacuation and preparation; building design: land-use change; permanent evacuation	Warning systems; emergency evacuation and preparation; building design; land-use change; permanent evacuation	Forecasting; rescheduling; inventory control; building design; seasonal adjustments (snow tires, chains); seasonal migration; designation of snow emergency routes
Adjust to losses:			
Spread the losses	Public relief; subsidized insurance	Public relief; subsidized insurance	Public relief; subsidized insurance
Plan for losses	Insurance and reserve funds	Insurance and reserve funds	Insurance and reserve funds
Bear the losses	Individual loss bearing	Individual loss bearing	Individual loss bearing

SOURCE: Ian Burton, Robert Kates, and Gilbert White, "The Human Ecology of Extreme Geophysical Events, "*Natural Hazard Research Working Paper No. 1,* Department of Geography, University of Toronto, 1968, p. 12. (This series is now published by the University of Colorado.) Reprinted by permission.

may produce an illusory feeling of security that encourages more rapid development of the flood plains and a relaxation of emergency preparations. This may lead to even greater damage when the next major flood occurs.

In most hazard studies inhabitants' impressions obtained by questionnaire surveys have been compared directly with the best physical measure available so that any discrepancies between the objective and subjective

measures of the same variable may be seen and evaluated. To explore further the relationships of the physical and subjective measures, geographers have used the strategy of selecting a series of study sites that vary along the most critical physical dimensions. Thus, for example, in studying drought, samples may be selected in zones ranging from mild to severe in terms of drought frequency.

In an early review of the entire range of hazards, Burton and Kates advanced a set of hypotheses that have since been substantiated in research on many natural hazards.[34] Variations in hazard perception were explained in terms of the relation of the hazard to the dominant resource use, the frequency of occurrence of the hazard, and variations in degree of personal experience. Heightened hazard perception is found where the hazard is directly related to the resource use. Thus agriculturalists are more aware of the flood hazard than urban dwellers, wheat farmers perceive the drought hazard more readily than cattle ranchers, and beach erosion is noticed at a waterfront cottage. The frequency of the natural events is related to the perception of the hazard. For example, in the Great Plains those in the most drought-prone area have the most accurate perceptions of the drought risk. Greater experience tends to lead to heightened hazard perception. This seems to hold true in flood plains, drought areas, and other hazard zones. Figure 6.1 illustrates the relationship between flood frequency and perception and adjustment for a number of urban places.

The perception of hazard—whether flood, drought, snow, landslides, tidal waves, or coastal storms—involves a high degree of abstraction and generalization of a complex reality. Although experts are able to work out with some precision the chances of recurrence of various hazards, they cannot, of course, predict exactly when they will occur in any one area. Nonprofessionals are even less at home with the probabilities of risk and with uncertainty. Commonly they try to eliminate the uncertainty by describing the hazards as cyclical phenomena or by taking a fatalistic view, saying it is out of their hands.[35] Another set of responses is to eliminate the hazard by denying its existence or recurrence. Table 6.4 provides some examples of these common responses.

Society's handling of extreme natural events might be considerably improved if the average person could be educated to think rationally about random events. When considered from a global perspective, so-called rare random events no longer appear infrequent or unusual. Kenneth Hewitt notes that the "once-in-two-thousand-years" event for a 10,000-square-mile area in North America may turn out to have an average global recurrence of several times per year.[36] If public officials could be persuaded to accept such facts, much loss of life and other damages could be avoided, on both national and world levels. Discussions could then take place to assess carefully the most desirable balance between the level of expected losses and the costs of further adjustments to avoid them. This would carry us beyond the current practice of responding only when a crisis occurs. At

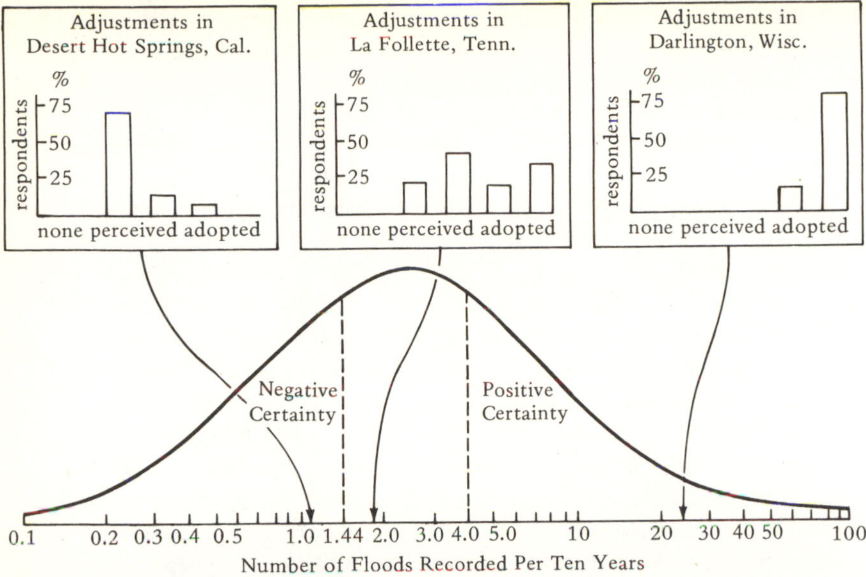

Fig. 6.1 Relationship between flood frequency and perception and adoption of adjustments. (Adapted from Ian Burton, Robert Kates, and Gilbert White, "The Human Ecology of Extreme Geophysical Events," *National Hazard Research Working Paper No. 1*, Department of Geography, University of Toronto, 1968, p. 20. This series is now published by the University of Colorado.)

such times there is a flurry of public works programs or studies, followed by a tendency to forget about the problem until the appearance of the next crisis.

The role of key decision-makers and their attitudes

Much attention has been devoted to the key decision-makers, whether these are individual resource managers, such as farmers, or public officials, whose attitudes may have a great influence. What is currently known about public attitudes toward the environment, some of the pitfalls involved in attitudinal studies, and how attitudes relate to action are discussed in a pair of papers by White and Lowenthal.[37] It is clear that attitudes are important factors in decision-making, though their precise role is not well worked out. They appear to enter into decisions in three ways: first, through the personal attitudes of those making decisions; second, through their opinions of what others prefer; and, third, through their opinions of what others should prefer. One problem is that too many decisions are made on the basis of someone's opinions of other people's attitudes without any real knowledge of what these attitudes are. That is one error that can be avoided by assessing public opinion. However, because such errors occur, it becomes important to probe the attitudes of the decision-makers.

Table 6.4 Common Responses to the Uncertainty of Natural Hazards

Eliminate the hazard		Eliminate the uncertainty	
Deny or denigrate its existence	Deny or denigrate its recurrence	Making it determinate and knowable	Transfer uncertainty to a higher power
"We have no floods here, only high water."	"Lighting never strikes twice in the same place."	"Seven years of great plenty . . . After them seven years of famine."	"It's in the hands of God."
"It can't happen here."	"It's a freak of nature."	"Floods come every five years."	"The government is taking care of it."

SOURCE: Ian Burton and Robert Kates, "The Perception of Natural Hazards in Resource Management." Reprinted with permission from 4 *Natural Resources Journal* 435 (1964) published by the University of New Mexico School of Law, Albuquerque, New Mexico.

Many of those who make decisions about the environment have a strong professional identification that tends to shape their perception of the environment. Thus economists, engineers, foresters, and others see environmental problems and solutions in terms of their professional role, and their views might be further narrowed to the principal concern of the agency that employs them.[38]

Elizabeth McMeiken found that public health officials in British Columbia perceived water quality problems in terms of potential impact on human health.[39] Comparison of these views with actual decisions or recommendations made by the officials in particular water quality situations indicated a close connection between such views and behavior. It is clear that the perceptions and attitudes of public health officials differ from the views of other groups as to the nature of the water quality problems, the solutions to be applied, and the appropriate role of various participants in the decision-making process. Yet many are loath to involve the general public or other agencies because they fear it would make decision-making cumbersome, inefficient, or perhaps even impossible.

Derrick Sewell extended the findings by comparing them with a similar study of engineers.[40] Engineers see themselves as technical advisers, not professionally competent to trace all the social impacts of particular decisions. Yet they are heavily relied upon for advice on a broad range of water problems with far-reaching social, political, and economic implications. Sewell found that

> the perceptions and attitudes of the two groups of professionals studied have all the characteristics of a closed system. Their views tend to be highly conditioned by training, adherence to standards and practices of the respective professions and allegiance to the agency's or firm's goals or mission.[41]

He concluded that changes will be needed in administrative structures, laws, and policies to ensure that a broader view is taken and that problems of environmental quality are considered in an integrated fashion with greater public participation.

Recent extensions of natural hazard research

The most recent work on perception of natural hazards could be considered a two-pronged attack on the problem of developing theory of greater generality. The logical extension of the research to a consideration of the perception of such artificial hazards as air and water pollution was followed by attempts to consider "all-hazards-at-a-place."[42] Instead of considering each hazard in isolation, there is an effort to examine the total hierarchy of environmental problems as they are perceived at a particular place. Meanwhile, other efforts were being made to test the type of questionnaire, developed mainly in North America, in a much broader range of study sites around the world.[43] This involves a large number of hazards and a wide variety of sites with people at different levels of economic development.

The "all-hazards-at-a-place" study design of Hewitt and Burton represents an extension of the ecological thinking of this research tradition. They consider hazards as extremes from the normal range of events to which communities are adjusted. While adjustments to normal conditions provide people with resources, the extreme events may cause severe deprivation or damages. To understand the responses to extreme events, one must understand the normal conditions and the thresholds that mark the limits of tolerance of the community. In studying all the hazards, both natural and artificial, that plague London, Ontario, the authors try to redefine hazards in terms of human response. Hewitt and Burton suggest that the magnitude and frequency of geophysical variables may not be the most suitable measures for defining an event's hazardousness for a particular community. In the case of snow storms, a major problem in Boston (Figure 6.2), the direct physical damage is secondary to the breakdown of power and communications. A more meaningful measure of the hazard impact would have to include the human response.

Similar in the emphasis on several problems in one place is a paper by Cooke and myself on public perception of environmental quality in Tucson, Arizona.[44] The paper presents the results of a survey conducted for the "Earth Day" Environmental Teach-In in Tucson. Residents were asked to rate many environmental problems in terms of seriousness, to rank in order the three most serious problems, and to provide their impressions on the causes and solutions for each. In this manner detailed information about public perception of specific problems was obtained, together with some idea of the relative importance of each in the community's hierarchy

Fig. 6.2 Heavy snow storm of February, 1969 Boston, Massachusetts. (J. Albertson, Stock, Boston.)

of problems. In Tucson a favorable image appeared dominated by the advantages of a fine climate. In ranking the problems, environmental and social issues alternated, as Table 6.5 shows. Comparative studies using some type of questionnaire might reveal how the dimension of perceived problems contributes to the uniqueness of different communities.

A major study of natural hazards in a wide variety of world sites was recently undertaken.[45] A large number of collaborators, all using a standard questionnaire, tested a series of general hypotheses on human adjustment to natural hazards. Sites in countries in various stages of economic development were included for each hazard. Thus, for example, there were studies of drought perception in Mexico, Nigeria, Brazil, Australia, and Tanzania; perception of air pollution in the United Kingdom, Yugoslavia, and Hungary; perception of the flood hazard at sites in the United States, Great Britain, Japan, and Ceylon; and perception of tropical cyclones in the United States and Bangladesh. Direct comparison of perceptions and adjustments to hazards may be possible, such as Table 6.6, prepared by Kates, which compares the views of Tanzanian hoe cultivators and American grain farmers.[46] It is apparent that the Americans see a wider range of choice in adjustments, including more related to farm practices and more requiring high-level technological input, such as construction of dams, ponds, terraces, and irrigation. In contrast, the Tanzanians seem more flexible in terms of changing their life pattern, and "the Tanzanian farmer seems willing to move with an uncertain nature, his American

Table 6.5 Ranking of Problems

Problem	First rank number	%	Second rank number	%	Third rank number	%	TOTAL number	% (*n*=202)*
Air pollution	46	23	26	13	15	7	87	43
Juvenile delinquency	26	13	22	11	15	7	63	31
Littering landscape	13	6	21	10	20	10	54	27
Traffic congestion	17	8	15	7	14	7	46	23
Noise pollution	17	8	11	5	15	7	43	21
Poor housing	20	10	14	7	7	3	41	20
Falling water table	15	7	13	6	10	5	38	19
Ugly streets	11	5	12	6	11	5	34	17
Urban sprawl	7	3	0	0	0	0	7	3
Soil erosion	4	2	1	0	2	1	7	3
Race problems	6	3	0	0	0	0	6	3
Water pollution	4	2	1	0	0	0	5	2
Lack of space for recreation	4	2	0	0	0	0	4	2
Labor problems	4	2	0	0	0	0	4	2
Overuse of pesticides	2	2	2	1	0	0	4	2
Lack of urban amenities	2	1	0	0	0	0	2	1
Interference with landowners	2	1	0	0	0	0	2	1
Arroyo cutting	0-	0	1	0	0	0	1	0
Lack of public housing	1	0	0	0	0	0	1	0
Lack of public transportation	0	0	0	0	0	0	1	0
Cotton prices	0	0	0	0	1	0	1	0
TOTAL	202		139		110		451	

*This total includes only those who attempted to rank the problems. SOURCE: Thomas Saarinen and Ronald U. Cooke, "Public Perception of Environmental Quality in Tucson, Arizona," *The Journal of Arizona Academy of Science*, 6 (1971), p. 265. Reprinted by permission.

counterpart appears ready to battle it out from a fixed site."[47] The validity of such conclusions may be disputed, since the data are subject to all the hazards of cross-cultural studies. One fundamental objection raised by students of cognitive anthropology is that use of a questionnaire is itself a questionable technique. The reason for this is that questionnaires force the interviewer's categories on the subject, who may think in totally different categories.

Cognitive anthropology

Cognitive anthropology represents a new theoretical orientation in anthropology that seeks to understand which material phenomena are significant for the people of a culture and how these phenomena are organized in their minds. "Where earlier anthropologists sought categories of description in their native language, cognitive anthropologists seek categories of description in the language of their natives."[48] Or, "cognitive anthropology entails an ethnographic technique which describes cultures from the inside out rather than from the outside in."[49]

The development of this field in the past decade has been impressive.

Table 6.6 Comparative Perception of Feasible Adjustment to Dought in Dry Areas of Tanzania (131 Farmers Interviewed) and the United States (96 Farmers Interviewed)

Tanzania			United States		
adjustments	**no.**	**%**	**adjustments**	**no.**	**%**
Q. If the rains fail, what can a man do?			*Q. If a meeting were held and you were asked to give suggestions for reducing drought losses, what would you say?*		
Do nothing, wait	17	12.14	No suggestions	16*	8.25
Rainmaking, prayer	15	10.71	Rainmaking, prayer	2	1.03
Move to seek land, work, food	51	36.43	Quit farming	1	0.52
Use stored food, saved money, sell cattle	16	11.43	Insurance, reserves, reduce expenditures, cattle,	16	8.25
Change crops	9	6.43	Adapted crops	2	1.03
Irrigation	15	10.71	Irrigation	46	23.71
Change plot location	4	2.86	Change land characteristics by dams, ponds, trees, terraces	26	13.40
Change time of planting	—	0.00	Optimum seeding date	—	0.00
Change cultivation methods	1	0.71	Cultivation: stubble mulch, summer fallow, minimum tillage, cover crops	78	40.21
Others	12	8.57	Others	7	13.61
TOTAL	140	99.99	TOTAL	194	100.00
Adjustments per farmer = 1.07			Adjustments per farmer = 2.02		

*Inferred from published report and subject to correction. SOURCE: Robert W. Kates, "Human Perceptions of the Environment," *International Social Science Journal*, 22 (1970), No. 4, 648-660.

There were forerunners, such as Malinowski who, in describing the garden activities of the Trobrianders, emphasized that the actual botanical facts do not matter so much as the process of growth seen through native eyes.[50] Now the anthropology journals fairly bristle with such terms as "formal analysis," "componential analysis," "folk taxonomy," "ethnoscience," "ethnosemantics," and "sociolinguistics." As many of the foregoing terms indicate, the main approach to the phenomenal worlds of the groups studied has been through language.

The researchers record how other people name all the things in their environment and how these are organized into larger groupings. "These names are thus both an index to what is significant in the environment of some other people, and a means of discovering how people organize their perceptions of environment."[51] Although the definitions and terms may vary, it is clear that, like most of the other researchers considered in this book, the cognitive anthropologists are seeking mental phenomena as data. Their aim or orientation, held in common with the many other disciplines discussed in other chapters, has been described as "the discovery of the organizing principles used by individuals, cultures, and species in manipulating and adapting to their particular life-space."[52]

To discover such organizing principles in other cultures has required the development of new fieldwork techniques and methods of analysis, and so far much of the work has involved developing and elaborating specialized techniques. Use of standardized questionnaires is out, for the categorization has already been accomplished by the investigator when such a tool is used. Instead, the method of controlled eliciting is used. This method is designed not only to provide the ethnographer with the answers but also to assist in discovering the relevant questions for the culture under examination. In spite of the striking parallels in interests, aims, and stage of development, there is virtually no contact between cognitive anthropologists and workers in other disciplines interested in how people perceive their environment. They may be aware of each other's work, but there is little evidence of it, even in footnote references, let alone theories or techniques.

Several types of taxonomies studied by cognitive anthropologists overlap the fields of interest of those concerned with environmental perception, behavior, and planning. Jane O. Bright and William Bright tell us that:

> . . . the inland Yurok and the Karok, living on either side of the Klamath River, oriented themselves not to the apparent motion of the sun, which the Europeans use to define the terms "north, south, east, west," but to the direction of river flow. This is reflected in the Yurok and Karok terms for cardinal directions, words translatable as "upriver, downriver, towards the river, away from the river.[53]

Einar Haugen explains the semantics of Icelandic orientation.[54] Both of these fall squarely in the realm of spatial orientation or cognitive mapping.

A favorite topic of cognitive anthropologists is that of folk taxonomies of the plants or animals within a particular culture region. Brent Berlin, Dennis E. Breedlove, and Peter H. Raven, for example, show the strong correlation between the cultural significance of species and the degree of linguistic differentiation of plants.[55] In comparing Tzeltal taxonomy with the standard botanical classification, they found three categories, as shown in Table 6.7. The first category consisted of plants that were underdifferentiated—i.e., Tzeltal terms included two or more botanical species. The second category consisted of terms with a one-to-one relation with botanical categories. The third category included overdifferentiated species, those for which more than one Tzeltal name corresponds to each botanical specific name. To explain the different lexical treatment accorded various plants, they were classified in terms of cultural significance. As shown in Table 6.8, there is a strong correlation between cultural significance and degree of lexical differentiation. Thus, for example, the category of high cultural significance includes all the plants intensively cultivated by the Tzeltal, such as, corn, beans, chili, and squashes. These important plants tend to be overdifferentiated. One might expect then, in general, that people will be much more discriminating in their perception of items

Table 6.7 Examples of the Three Categories of Tzeltal Specific Plant Names

Tzeltal specific name	Botanical classification
	Underdifferentiation
?ahate?es	*Archibaccharis flexilis* Blake (Compositae) *Gaultheria odorata* Willd. (Ericaceae) *Ugni montana* (Benth.) Berg (Myrtaceae) *Vaccinium leucanthum* C. & S. (Ericaceae)
?ičil?ak'	*Clematis dioica* L. (Ranunculaceae) *Clematis grossa* Benth. (Ranunculaceae) *Serjania* spp. (Sapindaceae)
	One-to-one correspondence
balamk'in	*Polymnia maculata* Cav. (Compositae)
kašlan bok	*Brassica oleracea* L. (Cruciferae)
¢a?te	*Ateleia pterocarpa* Sessé & Moc. (Leguminosae)
we?balil ¢ib	Marattia weinmaniifolia Liebm. (Marrattiaceae)
	Overdifferentiation
cahal šču?il čenek' *k'anal šču?il čenek'* *?ihk'al šlumil čenek'* *cahal šlumil čenek'* *sakil šlumil čenek'*	*Phaseolus vulgaris* L. (Leguminosae)
k'atk'at bohč *sepsep bohč* *¢u* *čahk'o?*	*Lagenaria siceraria* (Mol.) Standl. (Cucurbitaceae)

SOURCE: Brent Berlin, Dennis Breedlove, and Peter Raven, "Folk Taxonomies and Biological Classification," *Science*, 154 (October, 1966), 273-275. Reprinted by permission.

of importance to themselves. Studies by Harold Conklin show the direct relevance of such folk taxonomies in native concepts of ecology and in cropping patterns employed.[56]

Many well-meaning but ill-informed attempts by experts from industrialized nations to improve the agricultural techniques of people in the non-Western world have failed because they lacked awareness of such local folk taxonomies. It does not seem likely that in the future there will be fewer of these projects involving broad-scale human and landscape changes. However, it is to be hoped that the same sort of mistakes will not persist. Some can certainly be avoided by a greater understanding of the phenomenal world of the people concerned. The main dimensions of such a phenomenal world are clearly outlined by Ward Hunt Goodenough, a leader in the field of cognitive anthropology. His book, *Cooperation and Change*,[57] is an invaluable source for workers in cross-cultural situations or situations of rapid social change.

Table 6.8 Relation of Cultural Significance to Differentiation (in terms of Botanical Categories) of Tzeltal Specific Plant Names

Underdifferentiation	One-to-one correspondence	Overdifferentiation
	Low cultural significance	
49	10 (2)	5
	Moderate cultural significance	
31 (1)	31 (14)	5
	High cultural significance	
2	27 (24)	40

Numbers in parentheses indicate number of plants which were presumably introduced into Tenejapa after the Spanish conquest. SOURCE: Brent Berlin, Dennis Breedlove, and Peter Raven, "Folk Taxonomies and Biological Classification," *Science*, 154 (October, 1966), 273-275. Reprinted by permission.

Past perceptions of regions

The way the natural environment is perceived and used by various culture groups was the focus of the discussion on cognitive anthropology. The same focus on how aspects of the natural environment are perceived was seen in the studies on the perception of natural hazards. A recent development in historical geography has been the application of the perception approach to studies of people in past periods. Hugh Prince, in an excellent review of the subject, states it more strongly: "Perhaps the most important advance that has been made in historical geography in recent years has been a new view of the past as seen through the eyes of contemporary observers and a critical examination of the evaluations they made of the objects they observed."[58] Through a fresh understanding of how other people in other times have perceived reality, we may gain a fuller knowledge of the world in which we live.

Ideas about the Great Plains region in North America (Figure 6.3) have shifted dramatically during the short period of American approach and occupancy. The images of the plains held by Indians, fur traders, Spanish explorers, Mormons, surveyors, explorers, tourists, and settlers have all been explored, along with contrasts between the images held of the Canadian and American portions.[59] In the first half of the nineteenth century, when most Americans lived in the more humid eastern portion of the continent, the region was referred to as the Great American Desert. According to G. M. Lewis the present name, the Great Plains, came into general usage in the post-Civil War period, largely as a result of railroad propaganda designed to bring in settlers.[60] Figure 6.4 shows an example of such promotional literature. The desert image was gradually changed to fit one of the prevailing myths of the American West—the myth of an agricultural empire and associated imagery of the good life, which Henry Nash Smith calls the theme of the "Garden of the World."[61] He shows how this myth was accepted as fitting the Great Plains and had a decided effect on

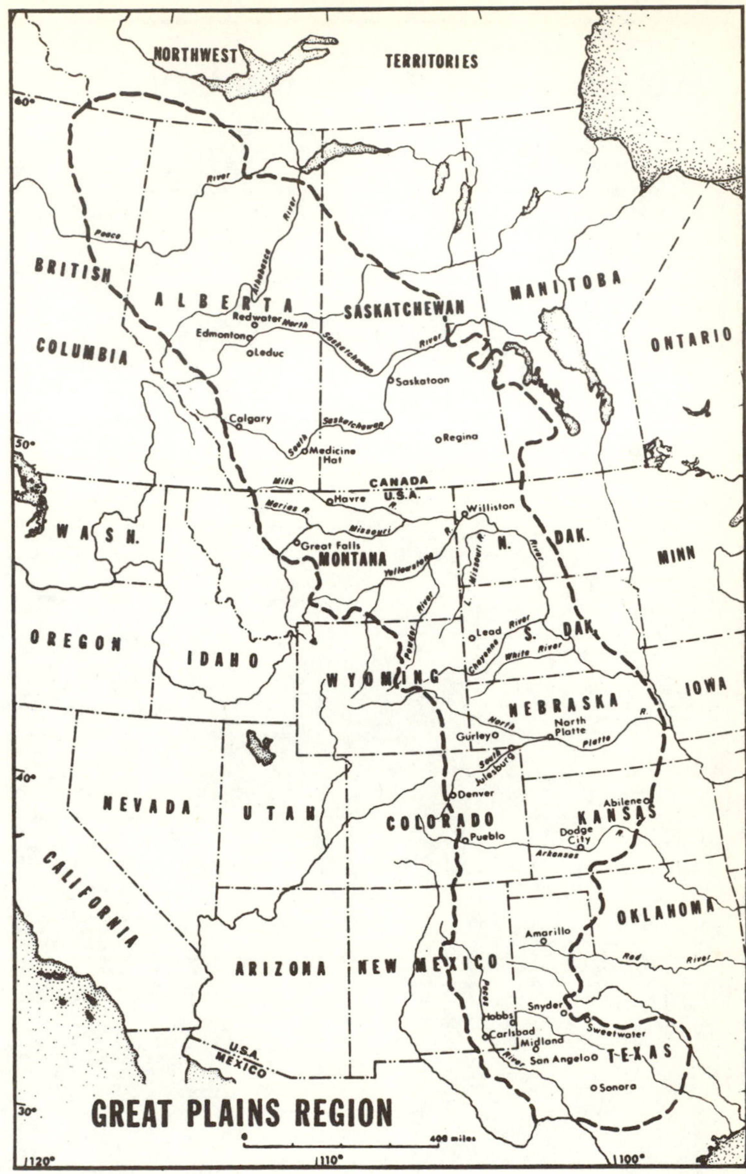

Fig. 6.3 Great Plains Region. (Paul F. Griffin, Ronald L. Chatham, and Robert N. Young, *Anglo-America: A Systematic and Regional Geography*, 2d. ed., Fearon Publishers, Belmont, California, 1968, p. 286. Reprinted by permission.)

the way people perceived it. Such erroneous dictums as "rain follows the plow" became widely accepted by both professional and lay people.[62]

Martyn Bowden examined the Great American Desert idea.[63] He

Actual Views in Morton County Kansas

These half-tones were made from actual photographs of the crops and products raised in Morton County, Kansas, last year (1912). They show the actual "juice," and affidavits are not necessary. Should anyone desire it we will send the name of owner and description of land upon which any one or all of these products were grown, or—come and we will show you.

United States Senator Thompson and Lieutenant-Governor Hopkins, both of Garden City, Kansas, looking over one of Morton County's milo maize fields. These distinguished gentlemen know the general conditions in southwest Kansas, and always speak a good word for Morton.

KANSAS

Kansas it seems to me, is not only the best pasture for all manner of beasts of the field, but it is the best field for rearing men and women. I know of no other place in civilization where a workingman's dollar will bring him so much comfort and so many luxuries, expose him to so few vices and surround him with such decent mental, moral and material influences that make for character and for every manner of prosperity, as in Kansas. The human crop of Kansas is the apex of the material pyramid which has for its base cattle, and corn, and alfalfa, and hogs, and wheat and prairie grass. Upon these foundations Kansas is building men whose ideals reach to the stars.

W. A[illegible]e, Editor Emporia Gazette.

A. J. Gerber, County Treasurer, in one corner of his hundred-acre field of milo, Kaffir, corn, sorghum, broom corn and water melons. Corn made 35 bushels per acre; maize, 45 bushels, and the seed from water melons net over $35 per acre. He is one of our progressive farmers.

STOCK RAISING

Morton county has always been considered by stock-men, on account of the wonderful supply of water and the excellent growth of native grass and the ease with which the different grains can be raised, to be one of the best places in the entire southwest for the raising of all classes of stock. As a rule cattle raising has predominated and many large ranches were located in Morton county, where thousands of high grade cattle were raised on the natural grasses. Horses, mules, sheep and hogs are all raised profitably, and in the past fifteen years there has been thousands of these animals shipped to eastern markets.

Our fine climate and the open winters reduces the expense of raising stock to the very lowest minimum, and many winters the stock has ranged entirely on the rich buffalo grass without any grain or hay whatever. Owing to excellent climatic conditions, all classes of stock thrive and very seldom there is a loss.

The large ranches have all been closed out, which gives the man of moderate means an opportunity to purchase a few quarters of our land and enter into the diversified farming which is being followed profitably by hundreds in the county at the present time, and with railroad facilities many fine opportunities are to be had.

The Santa Fe railroad is considered to be one of the greatest factors in the upbuilding of a country, and the fact that this company has purchased practically 90,000 acres of land in Morton county and is building a road into the county is conclusive evidence that the country is bound to advance. After the purchase of this land by the company it is placed on the market at very reasonable prices and terms, which affords the best opportunity ever offered in southwest Kansas for the home builder. All questions along any line pertaining to the county generally will be answered.

Alfalfa field under the large artesian well, showing one of the main ditches. This alfalfa produced almost five tons to the acre last year. This crop was sowed June 20, 1911.

MORTON COUNTY

Morton County is the southwest corner county of Kansas. Owing to the location it is in one of the very best climates in the entire southwest. Very seldom there are severe storms and the winters have been known to continue without any storms whatever. This certainly makes an ideal place in which to live.

One of the very best points in favor of the county is the exceptionally healthful climate. Many people who have come here for their health have regained it from the natural conditions without medicine.

The county has a forty thousand dollar court house, all paid for, and the good management of the county in the past has kept the taxes down within reason. The taxes range from $4.50 to $6.00 per quarter section, and the levy on personal property is very low.

There is an inexhaustible supply of shallow sheet-water throughout the entire county, and a wonderful flow of artesian water has been demonstrated. The county, according to government reports, is in one of the greatest artesian water basins known. These two points being proven certainly assures the future of the county.

The rain fall, according to carefully prepared observations for over twenty-five years show that from 16 to 25 inches fall every year, therefore it will be seen that but very little water outside of this is needed to make any crop.

The Santa Fe Railroad company now has the grade through the entire county completed, which will afford the best of transportation for all crops grown. It will certainly be of interest to all who contemplate purchasing good land to come and investigate and become conversant with the great opportunities which are offered along many of the different lines. The time to come is when the country is building.

One of the white cane fields one mile north of Richfield. The seed alone net over $15 per acre besides the fodder for stock. This is an easy and a sure crop for our county.

SANTA FE LANDS

The plat on other side shows all lands owned by the company which are for sale, at reasonable prices and exceptionally good terms. Prices range from $1600 to $2600 per quarter section, and terms are: one-eighth paid at time of purchase, balance in seven equal payments, last payment falling due at the end of the eighth year; second payment due at the end of the second year. Six per cent interest will be charged, and any part or all can be paid at any time and unaccrued interest rebated.

All particulars cheerfully given to anyone interested.

A fifteen-acre field of watermelons grown by Theodore Lewis. The seed from the melons netted over $40 per acre. How is this for $15 an acre land? The large seed houses know what can be done.

Exclusive Agents for Santa Fe Lands and Town Lots in Morton County.

WILSON & DEAN

RICHFIELD, KANSAS.

Fig. 6.4 Sample of Great Plains promotional literature. (Mel Hecht Collection, University of Arizona. Courtesy of Professor Mel Hecht.)

analyzed newspapers, atlases, gazetteers, geographies, and school books of the period 1800 to 1880 in an attempt to discover who believed in the Great American Desert, in which regions the idea was strong, when, and among which elements of the population. If such an image existed, it might well have helped curb migration to the Western Interior before the Civil War. Bowden found that the desert image was essentially the result of the Long expedition of 1819-1820. The idea was first disseminated to the educated elite of New England in the 1830s and later to the elite of the Northeast. He doubted that the image was shared by the elite of the South, by the Interior itself, or by the less educated people who were the potential settlers. He concluded that "the myth of the Great American Desert as the popular American image of the Western Interior before the Civil War is itself a myth."[64] Bowden's study demonstrates that the images of the elite group were entirely different from those of the general population. Furthermore, it is the latter group whose views were more important, as they were the potential settlers.

The Great Plains region is not unique in having a succession of changing images during different periods. Notions true and false have been documented for a variety of American regions, as, for example, the idea of insalubrious California,[65] the myth of a natural prairie belt in Alabama,[66] the error, started by Thomas Jefferson, that the climate of the Ohio country must resemble that of the southern seaboard,[67] and the view of South Carolina's physical environment as a terrestrial paradise in early promotional literature.[68] Every region occupied or known about over a long period is bound to have been perceived differently by different generations of inhabitants.

Robert M. Newcomb provides an illustration of this for a small area in Northern Jutland, Denmark.[69] He speculated on the ways people apparently perceived the Vester Han Herred district and how aspects of these perceptions were fulfilled in terms of natural resource exploitation. Five phases are described: the Stone Age mining of flint; the selection by Vikings of building sites for both ritual and settlement purposes; the medieval construction of parish churches out of glacial debris; the replanning of the rural landscape by means of enclosure; and the more recent impact of reclamation projects. Each of these phases involved a precise perception of particular resources, and each has left clearly demonstrable effects on the present-day land surface. All the resources were latent in the landscape from the Stone Age onward. The inhabitants of each period made their own appraisal of the resources available and selected for intensive development the resource they perceived as most appropriate in terms of their knowledge and needs. The same sort of cultural appraisal takes place in every society. An interesting change in what is perceived as a resource is the current concern in North America with areas suitable for various types of recreation.

Perception of wilderness

An expanding area of research in resource management is that concerned with how people perceive and use outdoor recreation areas. This might be expected to continue to grow as affluence, leisure time, mobility, and population increase and produce further pressure on outdoor recreation areas. As national parks and forests become more congested, isolated areas will become rarer and concern for wilderness preservation will grow. However, this concern may not be shared by all. Some people seek some sort of communion with nature and the privacy of the wilds, but many others prefer maximum social contacts and conveniences of home when they go camping.[70] The sociologists Gordon L. Bultena and Marvin J. Taves found five different images of wilderness among vacationers in Quetico-Superior in the Ontario-Minnesota border area:[71] (1) Wilderness as a locale for sport and play; (2) wilderness as fascination ("summons to adventure," "an opportunity to struggle with the elements"); (3) wilderness as sanctuary; (4) wilderness as heritage; and (5) wilderness as personal gratification. Alexander Grinstein, who examined vacations from a psychoanalytic framework,[72] considered the need for vacations to be "an expression of a very basic tendency which we may observe in the general cyclic character of the individual's life, as well as in the cyclic character of all living matter."[73] The important thing is to have some relaxation from cultural demands. This relaxation can occur within the framework of a continuous round of activity as well as in solitude, depending on the needs and preferences of the individual.

The need to study the preference patterns of individuals and groups to learn the directions and amounts of the total demand for recreation has been recognized by the Outdoor Recreation Resources Review Commission (ORRRC), which in 1960 carried out a massive program of surveys in 24 recreation areas in the United States. Among the series of resulting reports is one that assesses the quality of outdoor recreation on the basis of the perceptions of close to 11,000 respondents.[74] The major activities engaged in (Table 6.9) and the principal sources of satisfaction and dissatisfaction noted here have served as guidelines for managers of recreation areas and provided starting points for further perception studies. The results make it clear that there are many different types of groups seeking some form of outdoor recreation and that each has a characteristic array of preferred activities. In addition, certain types of areas seem more suitable for specific activities than others. The opinion survey in this context "provides an effective method of periodic evaluation of policy and program . . . as a guide for their reformulation in the light of current public desires and the capacity of the resource."[75] However, the public may perceive things differently from the recreation managers, as the ORRRC report noted in providing examples of areas considered overused by the

Table 6.9 Major Activities Engaged in by User Groups

Activity	Percent of groups in activity	Number of times ranked as one of five most commonly engaged in activities for an individual area					
		1st	2d	3d	4th	5th	1st-5th
1 Relaxing	53.1	1	5	6	6	4	22
2 Picnicking	51.1	3	4	5	5	-	19
3 Swimming	43.2	9	2	-	3	-	14
4 Sightseeing with stops	41.3	4	1	-	3	3	11
5 Walking to scenic points	39.9	1	2	5	4	-	12
6 Photography	39.3	-	3	3	-	1	7
7 Sunbathing	31.4	-	5	-	2	1	8
8 Camping	29.3	2	2	-	3	-	7
9 Sightseeing from car	26.2	-	-	-	-	1	1
10 Trail hiking	23.9	-	-	-	1	2	3
11 Wading	22.3	-	-	1	-	2	3
12 Visiting museums	14.2	-	-	-	-	2	2
13 Bank fishing	14.0	1	-	1	-	-	2
14 Looking at or collecting plants, animals, minerals	13.6	-	-	-	-	-	-
15 Listening to ranger talks	10.4	-	-	-	-	-	-
16 Boat fishing	9.4	-	-	-	-	2	2
17 Motorboating	9.3	-	2	-	-	3	
18 Snow skiing	7.0	2	-	-	-	-	2
19 Horseback riding	6.3	-	-	-	-	1	1
20 Rowboating	6.0	-	-	-	-	-	-
21 Fishing (wading)	5.9	-	-	-	-	-	-
22 Games and team sports	5.5	-	-	-	-	-	-
23 Water skiing	5.4	-	-	1	-	-	1
24 Taking guided tours	5.0	-	-	-	-	-	-

SOURCE: Outdoor Recreation Review Commission, "The Quality of Outdoor Recreation as Evidenced by User Satisfaction," *ORRRC Study Report No. 5*, U.S. Government Printing Office, Washington, D.C., 1962.

experts yet entirely satisfactory to most users. Another problem in maximizing area quality is the resolution of conflicts arising from incompatible uses of a recreational area. The authors of the report cite the example of Glacier National Park, where visitors desire to engage in water skiing and to use outboard motors, which are incompatible with the functions and purposes for which this park was created.

A technique applied by Robert C. Lucas indicates how perception studies may aid managers of wilderness areas in finding solutions to such problems.[76] He asked various park users interviewed in the Boundary Waters Canoe Area of northeastern Minnesota to delimit on a map of the area that portion they considered wilderness. Summary maps were drawn for each user type. Visitors differed among themselves and with the resource managers. Figure 6.5 shows some of the differences in area considered wilderness. Those using motorboats were less bothered by the presence of roads, crowding, or noise than were canoeists. The canoeists had the most restricted area and could be considered closer to the wilder-

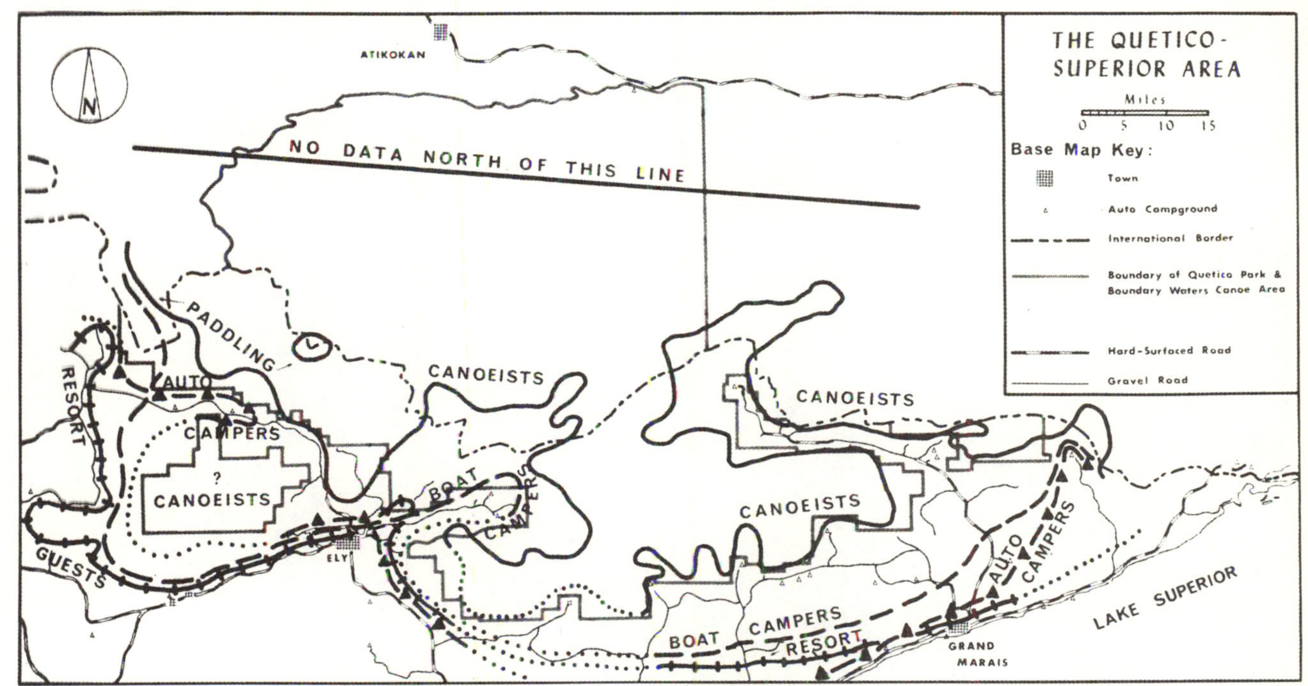

Fig. 6.5 The areas considered "wilderness" by at least 50 percent of the visitors in each of the four major types of users. The dotted portions of the lines indicate data were lacking and subjective estimates have been made based on 1960 data. (Robert C. Lucas, "Wilderness Perception and Use," in *Readings in Resource Management and Conservation*. eds. Ian Burton and Robert Kates, University of Chicago Press, © Chicago, 1965, p. 371. The University Of Chicago. Reprinted by permission.)

ness purist type. But, surprisingly, even for them, light logging did not seem incompatible with wilderness, nor was remoteness necessary if use was light. This parallels the finding of Lewis in a desert wilderness area where cattle grazing was not considered out of place in the wilderness—perhaps because of romantic associations with the Old West.[77] Lucas suggests that, by bearing in mind the different user definitions of wilderness, it may be possible to zone parks to provide a diversity of areas suitable for particular types of experience. Special attention would have to be paid to the perceptions of the more sensitive types of users, whose wilderness could most easily vanish through overuse. The paddling purist in the Boundary Waters Canoe Area would have his counterpart in the "backpacker" of the western mountain wildernesses. Various studies have sought to define the characteristics of such wilderness seekers and the conditions they consider essential for the wilderness experience.[78]

The conventional economic analysis, which estimates the monetary value of resources, fails to measure adequately the broad social values of such intangibles as scenery, relaxation, and release from urban stress provided by outdoor recreation. Studies of the perceptions and behavior of the recreation users provide better measures. As Lucas' work indicates, there appear to be differences in perception depending on the travel mode. These correspond to the differences in perception of cities according to travel mode noted in the previous chapter. Our perceptions of distance and space are functions of speed and time. Edward Abbey suggests the elimination of all motor traffic from national parks.[79] By so doing, the parks would seem far bigger than they do now, and the quality of the wilderness experience would be improved. In addition to travel mode, there are individual and group differences in how wilderness and other recreation areas are perceived and used. Recognition of such individual differences in environmental perception and behavior has led to an interest in defining the personality dimensions that might explain them.

Personality as a factor in environmental perception and behavior

The importance of personality differences as a factor in environmental perception and behavior has been recognized, and an increasing number of studies have focused on this theme. In the hazard studies, the attempt to measure aspects of personality directly related to resource use and adjustment to particular environmental circumstances has led geographers to use psychological techniques never previously applied in geographical research. Thus one can read of the results of the use of a modified version of the Thematic Apperception Test (TAT) among Great Plains wheat farmers in a work of mine[80] and another study by Sims and me,[81] of an adaptation of the Rosenzweig Picture-Frustration Test by Mary Barker and Ian Burton,[82] and of the use of a sentence completion test by John H. Sims and Duane

Baumann[83] to contrast the coping styles of people in Alabama and Illinois to the tornado threat. Here would seem to be a prime area for interdisciplinary cooperation between geographers and psychologists. The geographer, in his attempt to measure spatial variation in environmental ideas, usually selects samples from a series of different places. Such dimensions are rarely considered by the psychologist, who tends to select samples along psychological dimensions likely to be overlooked by geographers. Collaboration between the two could lead to fascinating results not anticipated by either discipline. Currently only crude measures of environmental personality are possible. Since it seems likely that the topic of environmental personality traits will become of even more importance in the future, we will examine briefly some of the starts that have been made.

Joseph Sonnenfeld, a geographer, has spent several years conducting cross-cultural research on environmental perception and sensitivity as well as on the role of personality factors as they may affect differential response to the physical environment. In comparing Arctic Alaskan and midlatitude Delaware populations, he developed a photo-slide test for investigating subjects' preferences for broadly differing types of landscapes.[84] Some fifty pairs of slides are used. The landscapes depicted vary according to four basic dimensions: topography or land forms, water, vegetation, and temperature. The slides are paired so that the subject must indicate preference for greater or lesser relief, richer or poorer vegetation, more or less water, and warmer or cooler temperatures. Since the test consists of slides, it has the advantage of being useful cross-culturally, as among the populations Sonnenfeld tested.

The tendency to prefer the home type of environment often accounted for some of the broad differences between populations. But there were differences within any one population as well. Certain people preferred the exotic or alien landscapes, which appeared attractive simply because they were different. The photo-slide test indicates only preferences, but, when a semantic differential technique was used to get at the meanings for choice of environmental elements, the same sort of intra-population differences were found.

Sonnenfeld suggested that certain preferences and attitudes toward the environment may prove universal. He went further to say, "I am willing to predict that consistent environmental personality types will be found among all populations, regardless of the contrast in cultural values otherwise distinguishing between them, and regardless of the contrast in environments they occupy."[85] Much of his later work has centered on the development of a personality inventory to measure such types.[86] His personality inventory as originally tested focused on four categories of behavior: or personal characteristics assumed to have implications for behavior: sensitivity to environment; mobility in environment; control over environment; and risk-taking in environment. Later these scales were recombined in a model of environmental personality, which includes, in

addition to the stimulus situation, personality variables grouped into categories of awareness, operational style, style of action/reaction, and motivation[87] (Figure 6.6).

In his comprehensive review of environmental psychology, Kenneth H. Craik devotes a section to discussion of personality inventories.[88] He points out the remarkable neglect by psychologists of items or scales for assessing environmental dispositions, in spite of the evidence that people display enduring orientations toward the physical environment. He suggests a number of environmental dispositions, such as a Pastoralism scale, an Ecological Perspective scale, a Luddite scale, an Urbanite scale, and measures of environmental sensitivity. George E. McKechnie has developed the Environmental Response Inventory (ERI) for assessing environmental dispositions,[89] which he defines as "the configuration of attitudes, beliefs, values, and sentiments" of the people being tested. Items were written to tap the themes of pastoralism, conservation, science and technology, urban life, rural life, stimulus preferences, cultural life, leisure activities, the outdoors, geographic and architectural preferences, and environmental memories and knowledge. The responses to these were factor-analyzed to reveal a series of separate factors for men and women. Further testing with other psychological tests of recognized usefulness enabled McKechnie to provide brief personality sketches of the persons scoring high on each scale. For example:

The person defined by a high score on the *Environmental Security* scale appears to have an environmentally restricted biographical history. She more likely than not

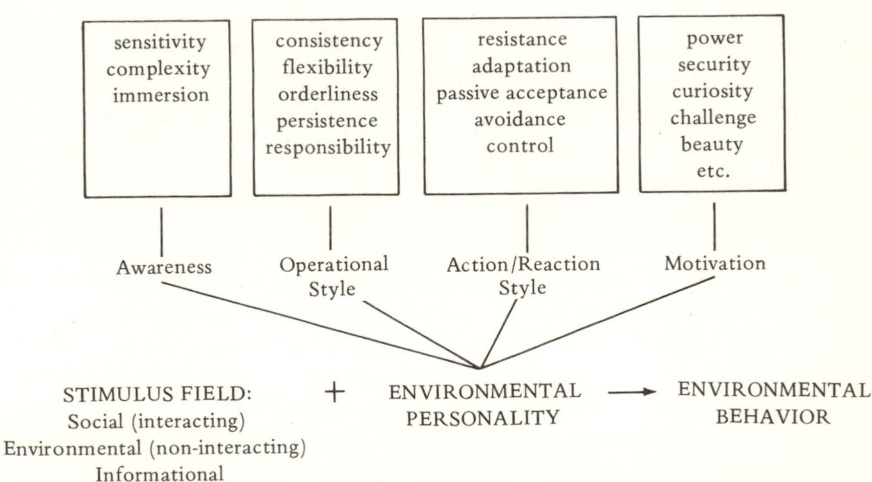

Fig. 6.6 A model of environmental personality and behavior (From Joseph Sonnenfeld, "Social Interaction and Environmental Relationship." Adapted from *Environment and Behavior,* Vol. 4, No. 3 (September, 1972), p. 268, by permission of the publisher, Sage Publications, Inc.)

comes from a rural, non-professional background. She scores generally low on measures of ascendancy, poise and self-assurance. She is somewhat conventional, unambitious, unassuming and lacking in self-confidence.[90]

In contrast:

> The male *Urbanism* scale appears to be positively related to social ascendancy, flexibility, and resourcefulness. The high-urban male likes the city and knows how to enjoy it. He is a dominant person but not pushy. His ascendancy is mediated by tact and consideration for others. He is, in a word, charming.[91]

Although the ERI is still being evaluated, many possible uses are foreseen, as, for example, relating the environmental types to different recreation preferences or using specific scales to predict frequency of wilderness use or urban adjustment.

Human dimensions of weather and climate

Most of the research discussed in this chapter has been concerned with human adaptation to the natural environment. The personality inventories, for example, are attempts to devise measures of environmental sensitivity. The wilderness studies suggest that many human needs may be satisfied through close contact with nature. The hazard studies indicate the pervasive power of natural events. Such inquiries raise the broader philosophical questions of climatic influences and environmental determinism. These include both the effect of the environment on people and the effect of people on the environment, both of which have long, rich traditions of discussion and disputation. Both are pertinent to the study of environmental perception and behavior. Some of these ideas will be discussed in relation to broad attitudes of Western civilization in Chapter 8. Here we will just note in passing a few recent developments, apart from the study of hazards, that provide impetus for research on how we perceive and adjust to weather and climate.

For some of the flavor of scientific experiments that link specific aspects of weather and climate directly to physiological reactions, there is no better single source than G. W. Tromp's mammoth compendium *Medical Biometeorology*.[92] It is notable for both the rigor with which the interactions are traced and the range of variables explored. The scope may be indicated by the number of factors treated. The physical factors included are those commonly considered, such as temperature, humidity, wind, and air pressure, as well as more subtle factors like electrical and magnetic properties of the atmosphere. In addition, many chemical, physicochemical, and air-polluting substances, and even the possibility of extraterrestrial factors are examined for possible influences on human beings. Regional differences in climate between countries and between areas in the

same country and their effect on human behavior and pathology are easier to demonstrate than daily weather influences. Nevertheless, Tromp has also gathered evidence available on the human biometeorological effects of microclimates and short-term weather changes in cities, inside buildings, and in space vehicles. An illustration of the pervasiveness of the influence of weather is provided by a study at the University of California. This research demonstrated that the electrical charge in the air within a newly constructed campus building changed with the broader weather patterns outside.[93] Such influences of the natural environment should be taken into account at all levels of planning. Some broad cultural responses to climatic variables that show up in variable house forms are discussed by Amos Rapoport in *House Form and Culture*.[94]

Current technology makes certain types of weather modification feasible. It is now possible to increase precipitation from winter or orographic storms, to disperse cold fog, and to suppress lightning. The possibility of large-scale applications of weather modification in the future has many social implications. These are explored in a symposium volume, on *Human Dimensions of Weather Modification*.[95] An increase in rainfall, for example, could be beneficial for wheat farming while harmful to irrigated agriculture in the same area. Many questions remain to be answered about the linkages to those downwind or downstream. Who would gain and who would lose? Who should make the decision to implement a weather modification program? Some of the same considerations together with concern over our increasing use of the atmosphere for waste disposal prompted a later volume, *Human Dimensions of the Atmosphere*.[96] The discussion here includes, among many other things, the effects of weather and climate on agriculture, on industry, on tertiary activities, and on other activities. W. J. Maunder discussed some of these effects in much greater detail in his book *The Value of Weather*,[97] which attempts "to bring together for the first time, the most significant and pertinent associations between man's economic and social activities, and the variations in his atmospheric environment."[98]

Summary

Some of the ideas discussed in this chapter illustrate features of environmental perception and behavior relevant at several scales. Although the scale is now getting beyond the possibility of immediate sense perception or single comprehensive views, some of the same principles still apply. The emphasis on extremes, first noted in the chapter on personal space, continues at the regional scale, where environmental hazards and wilderness areas are discussed. The concern with the views of the main decision-makers is expressed in clearer form here than in most of the other chapters, though its importance is not limited to the regional scale. In the hazard studies the perceived importance of considering the user groups con-

tinued. But again it became apparent that the people most affected are not generally consulted. New examples of the use of cognitive maps appeared, as in Lucas' study of wilderness perception and in the work of the cognitive anthropologists. In each chapter so far there have been psychological studies focusing mainly on the individual as well as anthropological studies emphasizing the importance of cultural factors. In this chapter these perspectives again emerged in the discussions of environmental personality and cognitive anthropology. The question of physical determinism, raised in connection with site planning, corresponds to environmental determinism in relation to climatic influences.

The emphasis on extremes as a focus for research was seen again in studies of natural hazards and wilderness areas. But in both cases the use of extreme cases may serve to highlight normal behavior. This idea was developed to some degree in the case of natural hazards, where the main stress of the research was indicated in terms of the broad ecological aim of our adjustment to our environment. Wilderness studies could also be thought of as one extreme in the total range of recreational choices. Since, at the present time, the growing demand for resources makes this particular, rather rare type more vulnerable than many others, it has, understandably, received a great deal of attention. But the same principles of identifying the users, determining their needs and preferences, and weighing these against the environmental possibilities apply at all scales, from a local "tot lot" to refuge areas for rare types of wildlife on a world level. They should also be valuable in planning and designing recreation for the specialized needs of different age groups, from infants to the elderly, and for special user groups, including snowmobile addicts, skiers, skin divers, white-water enthusiasts, and all others with specialized environmental demands.[99]

The emphasis on the decision-maker has always been a key factor in studies of resource management, and this has been developed more fully here than in the other chapters. Since decision-makers play a key role in environmental decisions, it becomes essential to find out who they are and what their conceptions are of their role and of the environment. For this reason it seemed appropriate to raise the question of environmental personality in this context. The regional level also seemed appropriate, as personality inventories aimed at examining environmental dimensions would have to include items of broad regional contrast, such as, rural versus urban areas, the effects of weather and climate, and regional differences.

The environment as considered at the regional scale was no longer limited to the built environment and the social environment. Instead, the natural environment became the new focus of attention, as in the studies of natural hazards, wilderness areas, weather modification, and folk taxonomies of plants and animals. The dimensions of the environment considered were more conceptual than perceptual. Generalizations were

based on long observation in such cases as perceived frequency of floods, categories of native plants, and long-range climatic variables rather than daily weather differences.

Cognitive anthropology was included in the chapter at the regional scale, although its approaches could be applied at other scales as well. The lack of interconnections between the work on cognitive anthropology and other research on environmental perception and behavior was noted. Though extreme in the case considered, the same tendency toward isolation of specific research groups may be seen at many levels. A concerted effort will be required to break through the barriers isolating disciplines and research traditions from one another.

The relevance of the perception approach to historical geography and studies of other cultures was seen to reside in the remarkably varied ways of viewing the environment at different times and in different places. Focus on the user group appears to be a fruitful approach. This leads to a need for new ways to measure the phenomenal world, a concern that will be pursued in the following chapters.

NOTES

1. See, for example, Alexander Spoehr, "Cultural Differences in the Interpretation of Natural Resources," in *Man's Role in Changing the Face of the Earth,* W. L. Thomas, ed., University of Chicago Press, Chicago, 1956, pp. 93-102.
2. For a good discussion of regional consciousness, see Chapter 4, "Regional Units and Regional Systems," in *Regional Ecology: The Study of Man's Environment,* Robert E. Dickinson, John Wiley & Sons, Inc., New York, 1970.
3. H. C. Darby, "The Problem of Geographical Description," *Transactions and Papers of the Institute of British Geographers,* No. 30, (1962), 1-13.
4. See, for example, H. C. Darby, "The Regional Geography of Thomas Hardy's Wessex," *Geographical Review,* 38 (1948), 426-433; J. H. Paterson, "The Novelist and His Region: Scotland through the Eyes of Sir Walter Scott," *Scottish Geographical Magazine,* 81, No. 3 (December, 1965), 146-152.
5. E. W. Gilbert, "The Idea of the Region," *Geography,* 45 (July, 1960), 157-175.
6. Otis W. Coan and Richard G. Lillard, *America in Fiction,* Pacific Books, Palto Alto, Calif., 1967. A brief world coverage is included in *Handbook for Geography Teachers,* M. Long, ed., Methuen and Company, Ltd., London, 1964, pp. 475-526.
7. An interesting recent example is John Warkentin, "Southern Ontario: A View from the West," *The Canadian Geographer,* 10, No. 3 (1966), 157-171. For a general discussion see D. W. Meinig, "Environmental Appreciation: Localities as a Humane Art," *Western Humanities Review,* 25, No. 1 (Winter, 1971), 1-11.
8. See, for example, Kenneth Clark, *Landscape Into Art,* Beacon Press, Boston, 1961.
9. John Kirtland Wright, *The Geographical Lore of the Time of the Crusades,* American Geographical Society, New York, 1925; republished as a paperback (Dover Publications, Inc., New York, 1965); for discussion of the first plea, Martyn J. Bowden, "John Kirtland Wright 1891-1969," *Annals of the Association of American Geographers,* 60, No. 2 (June, 1970), 394-403; for the first plea see Wright, "The History of Geography: A Point of View," *AAG Annals,* 14, 192-201; and "A Plea for the History of Geography," *Isis,* 8, 477-491. The latter is reprinted as Chapter

1 in Wright, *Human Nature in Geography,* Harvard University Press, Cambridge, 1966.

10. Ralph H. Brown, *Mirror for Americans: Likeness of the Eastern Seaboard (1810),* American Geographical Society, New York, 1943.
11. Wright, "Terrae Incognitae: The Place of Imagination in Geography," Chapter 5 in Wright, *Human Nature in Geography,* Harvard University Press, Cambridge, 1966, p. 83, first published in *AAG Annals,* 37 (1947), 1-15.
12. William Kirk, "Historical Geography and the Concept of the Behavioral Environment," *Indian Geographical Journal: Silver Jubilee Edition 1951,* Indian Geographical Society, Madras, 1952. See also his later statement, "Problems of Geography," *Geography,* 47, Part 4 (1963), 357-371.
13. H. C. Brookfield, "On the Environment As Perceived," *Progress in Geography,* 1, 51-80.
14. David Lowenthal, "Geography, Experience and Imagination: Towards a Geographical Epistemology," *AAG Annals,* 51, No. 3 (September, 1961), 241-260. Also Bobbs-Merrill Reprint G-137.
15. Gilbert F. White, *Human Adjustment of Floods,* Department of Geography Research Paper No. 29, University of Chicago, Chicago, 1946.
16. Robert William Kates, *Hazard and Choice Perception in Flood Plain Management,* Department of Geography Research Paper No. 78, University of Chicago, Chicago, 1962, and White, *Choice of Adjustment to Floods,* Department of Geography Research Paper No. 93, University of Chicago, Chicago, 1964. See also White, "Optimal Flood Damage Management: Retrospect and Prospect," in Allen V. Kneese and Stephen C. Smith, eds., *Water Research,* Johns Hopkins Press, Baltimore, 1966, pp. 251-269.
17. Ian Burton and R. W. Kates, "The Perception of Natural Hazards in Resource Management," *Natural Resources Journal,* 3, No. 3 (January, 1964), 412-441.
18. Burton *et al., The Shores of Megalopolis: Coastal Occupance and Human Adjustment to Flood Hazard,* Vol. 18, No. 3, C. W. Thornthwaite Associates Laboratory of Climatology, Elmer, N.J., 1965.
19. Thomas F. Saarinen, *Perception of the Drought Hazard on the Great Plains,* Department of Geography Research Paper No. 106, University of Chicago, Chicago, 1966.
20. John F. Rooney, Jr., "The Urban Snow Hazard in the United States," *Geographical Review,* 57, No. 4 (October, 1967), 538-559. See also Duane D. Baumann and Clifford Russell, "Urban Snow Hazard: Economic and Social Implications," *University of Illinois Water Resources Research Report No. 37,* University of Illinois, Urbana, Ill., 1971.
21. J. G. Michael Parkes, "Awareness and Adjustment to a Natural Hazard: Sensitive Clays in the Ottawa Area" (Master's thesis, University of Western Ontario, 1969).
22. R. Havighurst, "Tsunami Perception in Selected Sites on Oahu" (Master's thesis, University of Hawaii, 1967).
23. David H. K. Amiran, "Frost Hazard in Intensive Agriculture in the Negev: Topoclimatic Investigations," paper presented at UNESCO Seminar on Natural Hazards, Godollo, Hungary, August 4-9, 1971.
24. M. Aminul Islam, "Human Adjustment to Cyclone Hazards: A Case Study of Char Jabbar," Natural Hazard Research Working Paper No. 18, University of Toronto, Department of Geography, Toronto, 1971.
25. Tapan Mukerjee, "Economic Analysis of Natural Hazards: A Preliminary Study of Adjustments to Earthquakes and Their Costs," Natural Hazard Research Working Paper No. 17 (Toronto: University of Toronto, Department of Geography, 1971); see also Committee on the Alaska Earthquake of the Division of the Earth

Sciences National Research Council, *The Great Alaska Earthquake of 1964* (Washington, D.C.: National Academy of Science, 1970).

26. White, "Formation and Role of Public Attitudes," in *Environmental Quality in a Growing Economy,* H. Jarrett, ed., Johns Hopkins Press, Baltimore, 1966, pp. 105-127; Lowenthal, "Assumptions behind Public Attitudes," in Jarrett, ed., *op. cit.,* pp. 128-137.
27. White and Fred L. Strodtbeck, eds., *Attitudes toward Water: An Interdisciplinary Exploration,* University of Chicago Social Psychology Laboratory and Department of Geography, Chicago,
28. National Science Foundation, *Human Dimensions of the Atmosphere,* U.S. Government Printing Office, Washington, D.C., 1968.
29. See, for example, D. D. Baumann, *The Recreational Use of Domestic Water Supply Reservoirs: Perception and Choice,* Department of Geography Research Paper No. 121, University of Chicago, Chicago, 1969, and Shue Tuck Wong, *Perception of Choice and Factors Affecting Industrial Water Supply Decisions in Northeastern Illinois,* Department of Geography Research Paper No. 117, University of Chicago, Chicago, 1969.
30. D. R. Lycan and W. R. D. Sewell, "Water and Air Pollution as Components of the Urban Environment of Victoria," *Geographical Perspectives,* Tantalus Press, Vancouver, B.C., 1968, pp. 13-18; J. Elizabeth McMeiken and John Rostron, "Perception of Pollution and Attitudes towards Its Solution: A Pilot Study in Courtney and Victoria, British Columbia," *Geographical Studies,* Department of Geography, University of Victoria, Victoria, B.C.,
31. Kenneth Hewitt and Ian Burton, *The Hazardousness of a Place: A Regional Ecology of Damaging Events,* Department of Geography Research Publication No. 6, University of Toronto, Toronto, 1971.
32. Gilbert F. White, ed., *Natural Hazards: Local, National, Global,* Oxford University Press, New York, 1974.
33. Burton, Kates, and White, *The Human Ecology of Extreme Geophysical Events,* Natural Hazard Research Working Paper No. 1, University of Toronto, Department of Geography, Toronto, 1968.
34. Burton and Kates, "The Perception of Natural Hazards in Resource Management," *Natural Resources Journal,* 3, No. 3 (January, 1964), 421-441.
35. *Ibid.,* p. 435.
36. Kenneth Hewitt, *A Pilot Survey of Global Natural Disasters of the Past Twenty Years,* Natural Hazard Research Working Paper No. 11, University of Toronto, Department of Geography, Toronto, 1969, p. 1.
37. See Footnote 26.
38. The effect of the agency is described in Hubert Marshall, "Politics and Efficiency in Water Development," in Kneese and Smith, eds., *Water Research,* Johns Hopkins Press, Baltimore, 1966, pp. 291-310.
39. McMeiken, "Public Health Professionals and Attitudes in Environmental Quality Decision-Making in British Columbia" (Master's thesis, University of Victoria, 1970).
40. W. R. D. Sewell, "Environmental Perceptions and Attitudes of Engineers and Public Health Officials," *Environment and Behavior,* 3, No. 1 (March, 1971), 23-59.
41. *Ibid.,* 40.
42. See Footnote 31.
43. Gilbert F. White, ed., *Natural Hazards: Local, National, Global,* Oxford University Press, New York, 1974.
44. Thomas F. Saarinen and Ronald U. Cooke, "Public Perception of Environmental Quality in Tucson, Arizona," *Journal of the Arizona Academy of Science,* 6 (June, 1971), 260-274.

45. See Footnote 43.
46. Kates, "Human Perception of the Environment," *International Social Science Journal,* 22, No. 4 (1970), 648-660.
47. *Ibid.,* p. 658.
48. Stephen A. Tyler, ed., *Cognitive Anthropology,* Holt, Rinehart and Winston, Inc., New York, 1969, p. 6.
49. *Ibid.,* p. 20.
50. Bronislaw Malinowski, *The Coral Gardens and Their Magic,* two vols., American Book Company, New York, Cincinnati and Chicago, 1935.
51. Tyler, *op. cit.,* p. 6.
52. *Ibid.,* p. 20.
53. Jane O. Bright and William Bright, "Semantic Structures in Northwestern California and the Sapir-Whorf Hypothesis," in Tyler, *op. cit.,* p. 69; originally published in *American Anthropologist,* 67, No. 5, Part 2 (Special Publication) (October, 1965), 249-258.
54. Einar Haugen, "The Semantics of Icelandic Orientation," in Tyler, *op. cit.,* pp. 330-342; originally published in *Words,* 13 (1957), 447-460.
55. Brent Berlin, Dennis E. Breedlove, and Peter H. Raven, "Folk Taxonomies and Biological Classification," in Tyler, *op. cit.,* pp. 60-66; originally published in *Science,* 154 (October 14, 1966), 273-275.
56. Harold Conklin, *Hanunoo Agriculture,* Forestry Development Paper No. 12, FAO of United Nations, Rome, 1957, and Conklin, "An Ethnoecological Approach to Shifting Agriculture," in Philip L. Wagner and Marvin W. Mikesell, eds., *Readings in Cultural Geography,* University of Chicago Press, Chicago, 1962, pp. 457-464.
57. Ward Hunt Goodenough, *Cooperation in Change,* Russell Sage Foundation, New York, 1963; see especially Chapter 10.
58. Hugh C. Prince, "Real, Imagined and Abstract Worlds of the Past," *Progress in Geography,* Vol. 3, Edward Arnold, London, 1971, pp. 1-86; quote on p. 24.
59. Brian W. Blouet and Merlin P. Lawson, *Images of the Plains,* University of Nebraska Press, Lincoln, Nebr. (forthcoming).
60. G. M. Lewis, "Changing Emphases in the Description of the Natural Environment of the American Great Plains Area," *Transactions and Papers, Institute of British Geographers,* 30 (1962), 75-90; see also Lewis, "Regional Ideas and Reality in the Cis-Rocky Mountain West," *Transactions . . . British Geographers,* 38 (1965), 135-150.
61. Henry Nash Smith, *Virgin Land, The American West as a Symbol and Myth,* Vintage Books, New York, 1950, pp. 138-305.
62. For a lively discussion of this, see Walter M. Kollmorgen, "Rainmakers on the Plains," *Scientific Monthly,* 40 (1935), 146-152; and a later paper with a broader perspective, Walter and Johanna Kollmorgen, "Landscape Meteorology in the Plains Area," *AAG Annals,* 63, No. 4 (December, 1973), 424-441.
63. Bowden, "The Perception of the Western Interior of the United States, 1810-1870: A Problem in Historical Geosophy," *Proceedings of the Association of American Geographers,* 1 (1969), 16-21.
64. *Ibid.,* 21.
65. Kenneth Thompson, "Insalubrious California: Perception and Reality," *AAG Annals,* 59, No. 1 (March, 1969), 50-64.
66. Erhard Rostlund, "The Myth of a Natural Prairie Belt in Alabama: An Interpretation of Historical Records," *AAG Annals,* 47, No. 4 (December, 1957), 392-411. Also Bobbs-Merrill Reprint G-198.
67. As noted in Brown, "The Treatment of Geographic Knowledge and Understanding in History Courses, with Special Reference to American History," in *Geographic Approaches to Social Education,* Nineteenth Yearbook of the National

Council for the Social Studies, Clyde F. Kohn, ed., The National Council for the Social Studies, Washington, D.C., 1948, pp. 262-272; quote on p. 271.

68. H. Roy Merrens, "The Physical Environment of Early America: Images and Image Makers in Colonial South Carolina," *Geographical Review*, 59, No. 4 (October, 1969), 530-556.
69. Robert M. Newcomb, "Environmental Perception and Its Fulfillment during Past Times in Northern Jutland, Denmark," *Skrifter Fra Geografisk Institut Ved Aarhus Universitet Nr. 26*, Geografisk Institut Aarhus Universitet, Aarhus, 1969.
70. For discussion of this theme, see Lowenthal, "Daniel Boone Is Dead," *Natural History*, 77, No. 7 (August-September, 1968), 8-16 and 64-67.
71. Gordon L. Bultena and Marvin J. Taves, "Changing Wilderness Images and Forestry Policy," *Journal of Forestry*, 59 (March, 1961), 161-171.
72. Alexander Grinstein, "Vacations: A Psycho-Analytic Study," *International Journal of Psychoanalysis*, 36 (1955), 177-185.
73. *Ibid.*, p. 177.
74. *The Quality of Outdoor Recreation: As Evidenced by User Satisfaction*, ORRRC Study Report No. 5, Outdoor Recreation Resources Review Commission, Washington, D.C., 1962.
75. *Ibid.*, p. 45.
76. Robert C. Lucas, "Wilderness Perception and Use: The Example of the Boundary Waters Canoe Area," *Natural Resources Journal*, 3, No. 3 (1963), 394-411; and Lucas, "The Contribution of Environmental Research to Wilderness Policy Decisions," in R. W. Kates and J. F. Wohlwill, "Man's Response to the Physical Environment," *Journal of Social Issues*, 22, No. 4 (October, 1966), 116-126.
77. Jeffrey W. Lewis, "Perception of Desert Wilderness by the Superstitious Wilderness User" (Master's thesis, University of Arizona, 1971).
78. See, for example, William R. Burch, Jr., "Wilderness—The Life Cycle and Forest Recreational Choice," *Journal of Forestry*, 64 (September, 1966), 606-610; L. C. Merriam and R. B. Ammons, "Wilderness Users and Management in Three Montana Areas," *Journal of Forestry*, 66, No. 5 (May, 1968), 390-395; William R. Catton, Jr., and John C. Hendee, "Wilderness Users: What Do They Think," American Forests (September, 1968), 28-31 and 61-62; Elwood L. Shafer, Jr., and James Mietz, "Aesthetic and Emotional Experiences Rate High with Northeast Wilderness Campers," *Environment and Behavior*, 1, No. 2 (December, 1969), 187-197.
79. Edward Abbey, *Desert Solitaire*, Balantine Books, New York, 1968.
80. See Footnote 19.
81. John H. Sims and Thomas F. Saarinen, "Coping with Environmental Threat: Great Plains Farmers and the Sudden Storm," *AAG Annals*, 59, No. 4 (December, 1969), 677-686.
82. Mary Barker and Ian Burton, *Differential Response to Stress in Natural and Social Environments: An Application of a Modified Rosenzweig Picture-Frustration Test*, Natural Hazard Working Paper No. 5, University of Toronto, Department of Geography, Toronto, 1969.
83. John H. Sims and Duane Baumann, "The Tornado Threat: Coping Styles of the North and South," paper prepared for UNESCO Seminar on Natural Hazards, Godollo, Hungary, August 4-9, 1971.
84. Joseph Sonnenfeld, "Variable Values in Space and Landscape: An Inquiry into the Nature of Environmental Necessity," *Journal of Social Issues*, 22, No. 4 (1966), 71-82; also Sonnenfeld, "Environmental Perception and Adaptation Level in the Arctic," in *Environmental Perception and Behavior*, Lowenthal, ed., Department of Geography Research Paper No. 109, University of Chicago, Chicago, 1967, pp. 42-59.
85. Sonnenfeld, "Equivalence and Distortion of the Perceptual Environment," *En-*

vironment and Behavior, 1, No. 1 (June, 1969), 83-99, quote on p. 97.

86. Sonnenfeld, "Personality and Behavior in Environment," *AAG Proceedings,* 1 (1969), 136-140; "Personality and Risk in Environment," paper presented at the Annual Meeting of the Southwest Social Science Association, Houston, Tex., April 5, 1969.
87. Sonnenfeld, "Behavioral Dimensions of Cultural Geography," prepared for Colloquium on New Departures in Theoretical Cultural Geography, Sixty-sixth Annual Meeting of the Association of American Geographers, San Francisco, August 24, 1970; see also Sonnenfeld, "Monadic and Dyadic Approaches to the Study of Behavior in Environment," *AAG Proceedings,* 3 (1971), 165-169.
88. Kenneth H. Craik, "Environmental Psychology," in *New Directions in Psychology,* Vol. 4, Holt, Rinehart and Winston, Inc., New York, 1970, pp. 1-121.
89. George E. McKechnie, "Measuring Environmental Dispositions with the Environmental Response Inventory," *EDRA Two,* Proceedings of the Second Annual Environmental Design Research Association Conference, Pittsburgh, Pennsylvania, October, 1970, pp. 320-326.
90. *Ibid.,* p. 325.
91. *Ibid.,* p. 324.
92. G. W. Tromp, *Medical Biometeorology: Weather, Climate and the Living Organism,* Elsevier Publishing Company, Amsterdam, 1963. See also the very readable account from the point of view of health and disease, René Dubos, *Man Adapting,* Yale University Press, New Haven, Conn., 1965.
93. As reported in Tromp, *op. cit.,* p. 82.
94. Amos Rapoport, *House Form and Culture,* Prentice-Hall, Inc., Englewood Cliffs, N.J., 1969.
95. Sewell, ed., *Human Dimensions of Weather Modification,* University of Chicago Department of Geography Research Paper No. 105, University of Chicago, Chicago, 1966. See also J. Eugene Haas, Keith S. Boggs, and E. J. Bonner, "Science, Technology and the Public: The Case of Planned Weather Modification," in *Social Behavior, Natural Resources and the Environment,* William R. Burch, Jr., Neil H. Cheek, Jr., and Lee Taylor, eds., Harper & Row, Publishers, New York, 1972, Chapter 7.
96. National Science Foundation, *Human Dimensions of the Atmosphere,* U.S. Government Printing Office, Washington, D.C., 1968. See also Sewell, Kates, and Phillips, "Human Response to Weather and Climate: Geographical Contributions," *Geographical Review,* 58, No. 2 (1968), 262-282.
97. W. J. Maunder, *The Value of Weather,* Methuen and Company, Ltd., London, 1970.
98. *Ibid.,* p. xxi.
99. For an excellent empathic discussion of the development of such diverse approaches to exploration of the world, see J. B. Jackson, "The Abstract World of the Hot-Rodder," *Landscape,* 7, No. 2 (Winter, 1957-1958), 22-27; also reprinted in the Bobbs-Merrill Reprint Series in Geography G-104.

Seven

The nation

"Landscapes are formed by landscape tastes. People in any country see their terrain through preferred and accustomed spectacles, and tend to make it over as they see it."
David Lowenthal and
Hugh C. Prince

The nation, as considered in this chapter, is essentially the segment of environment included within the boundaries of a political state. A nation may be considered a mosaic of many unique regions and culture groups. But those within a country may be expected to share, to varying degrees, a common set of characteristics that bind them together in the higher-order organization represented by the nation. To isolate the common set of characteristics requires a high degree of abstraction and generalization. Studies of environmental perception and behavior at the scale of the nation could be described most accurately as concerned with conceptions of environment, or with the images, attitudes, and ideas most commonly held by the citizens of the country. Many broad types of investigation would fit here. One might be an examination of the degree to which various subcultures share in a dominant national image or idea. Another might be research into individual or group conceptions of national regional structure. Still another might be the attempt to abstract a general set of traits common to the national culture and their implications for environmental decision-making. Common to all such studies is a concern with abstract characteristics of people and places rather than direct investigation of people's perceptions of specific portions of the environment.

We will begin with a discussion of how the inhabitants conceive various regions within the nation. But here the concern is not so much with the individual regions as with the total set of regions and how they fit together within the nation. Later sections deal with the more abstract qualities of the nation as a whole. The topics discussed are regional consciousness, studies of place preferences, place preferences and migration, landscapes and landscape tastes, national attitudes toward the environment, and national character.

Regional consciousness

Regional consciousness[1] expresses aspects of environmental perception and behavior we have already considered in Chapter 4 on the neighborhood. There we noted the strong sense of belonging to their neighborhood of Boston's West Enders. The same sort of loyalty or sense of shared existence is found at all scales, from small localities to countries or even world regions. It is derived from attitudes, origins, associations, and organizations. Names[2] given to particular regions to label the perceived similarity of people associated with them reveal its presence—i.e., Snob Hollow, Lower East Side, Calgarian, New Englander, Texan, Western Europe, Arab.

Regional consciousness has a hierarchical aspect, so that the same person may feel a sense of belonging to a neighborhood, a city, a region, a country, or even broader sections of the world. The farther away from home one is, the broader the regional unit with which one may tend to identify. Thus within our city we may feel a loyalty to our neighborhood, in a nearby city a consciousness of belonging to our own city, and in another part of our country a regional loyalty, while overseas we identify with our country. Even in the most remote portions of the world broad regional or cultural loyalties may be felt. This extended sense of shared fate seems to have been stimulated by the recent excursions of the astronauts. The photos of the earth taken from the moon may have been instrumental in the increasingly widespread references to the image of "Spaceship Earth" with all human beings as fellow passengers. The delimitation of planning regions represents a practical application of regional consciousness. Planners could deliberately enhance the expression of the unique qualities of place by more sensitively appraising regional identifications.

In long-settled regions, common local traditions and interests have led to a strong sense of regional consciousness. One example of this is Mauri Palomaki's investigation of the present-day provinces of Finland, which illustrates a close correspondence between professional and lay impressions of regions.[3] But regional consciousness also develops in areas with a shorter history of settlement. Ruth Hale's study of vernacular regions in the United States demonstrates the remarkable diversity of areas seen as dis-

tinct by the local inhabitants.[4] In the United States, however, there is less agreement between lay and professional views of regions.

In Finland, the province as a regional concept is well developed. People have a strong sense of belonging to particular provinces. Palomaki used this sense of regional consciousness to delimit the present-day provinces of Finland. A sample of over 5,000 primary-school teachers was asked, "to what province do you consider your school district belongs?"

The results of the survey were compared with previously defined functional provinces, as Figure 7.1 shows. The functional provinces were defined by the "central-place theory," which is based on the distribution of places providing goods and services of various types. Although there was in general a good agreement, certain differences appeared. Most outstanding is the presence in people's minds of several provinces in the north that have not as yet developed effective central places. The province of Oulu, for example, splits into three parts, while that of Rovaniemi divides in two. A redrawing of boundaries and provision of services currently lacking would serve to support the newly emerging perceptual provinces. A contrasting case is seen in the historical provinces of Savo and Häme, which illustrate enduring ties created by a common history. These remain conceptual entities though each is made up of more than one functional province.

A doctoral dissertation by Ruth Hale was also concerned with regional consciousness. But her approach was different. She attempted to contact at least three representative from each of the counties and parishes in the conterminous United States. Postcards were sent to county agricultural and extension agents, weekly newspaper editors and publishers, and postmasters. They were asked to name the region in which their county was located and the region of the nation in which their state was located. Simultaneously, geographers in each of the 48 states were asked to draw on their state map the outlines of popularly identified regions and to signify in which national region they believed their state was located.

Hale found that a rich and colorful variety of names were used to designate different regions. In total, 295 vernacular local regions were identified. The names used reflect the great variety of cultural, historical, and natural events and elements that have been noticed and experienced by the inhabitants. For example, in California, where regional consciousness was at a high level, seven separate subregions were commonly identified. Central Coast, Central Valley (San Joaquin), Sacramento Valley, and Sierra all are derived from physical features. Redwood Region shows an identification with the local vegetation, Mother Lode with a form of economic activity, and San Francisco Bay Area with a political region. Other names, such as Little Dixie and Boonslick in Missouri, are historical in origin.

There was no shortage of locally identified regions, but in many cases there was less than complete consensus. In several areas all the respondents concurred on a regional name, but in others local regional consciousness was not well developed. The same held true at the level of primary national

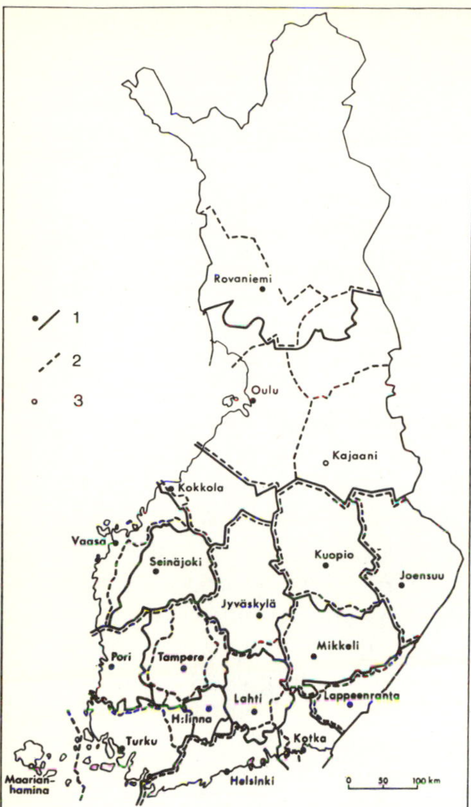

Fig. 7.1 Comparison of functional and conceptual provinces of Finland. (Mauri Palomaki, "On the Concept and Delimitation of the Present Day Provinces of Finland," *Acta Geographica,* No. 20, Geographical Society of Finland, p. 292. Reprinted by permission.)

region, as Table 7.1 shows. The Southwest emerged as the most strongly defined region. Over three quarters of the respondents there agreed that their state belonged in the southwest region. But within this broad region there was often a failure to identify smaller local regions. For example, in Arizona only three local regions were rather dimly perceived.

Figures 7.2 and 7.3 offer a comparison of the vernacular local regions and those perceived by geographers. Although many of the same names are used, the boundaries do not generally correspond except in the case of the Southwest. The geographer's South and New England are larger containing areas, which the other respondents referred to as the Southeast and Northeast. In the cases of the Pacific Coast, the Pacific Northwest, and the Midwest, the geographer's regions are less inclusive. Portions of these regions, as delineated on the map of vernacular regions, are given separate status. An interesting example is the Great Plains. It is distinguished by the geographers but included as part of the Midwest on the map of vernacular

Table 7.1 Consciousness of Primary National Region (Regional averages of state responses)

State regions	Percentage
Southwest	75.7
Pacific Northwest	66.6
(Far) West (Nevada)	64.0
Middle Atlantic	61.4
Midwest	60.2
Rocky Mountain States	58.3
New England (Connecticut)	58.0
South	54.2
Northeast	53.0
Southeast	50.1
Pacific Coast (California)	46.7

SOURCE: Ruth Hale, "Maps of Vernacular Regions in America," unpublished thesis, University of Minnesota, June, 1971, p. 71

regions. Carl Kraenzel has suggested that this lack of a clearly perceived regional identity is a major problem for the Great Plains.[5] He contends that it encourages a colonial status and impedes rational regional planning.

The rather close correspondence between Finnish professional and lay identifications of regions was not duplicated in the case of the United States. But in both cases it was seen that awareness of regional consciousness may provide insights that could aid in the delineation of planning regions corresponding to people's local loyalties. The rich variety of regions identified in the Hale study and the great differences in degree to which local regions were recognized raise many questions about the factors that influence the development of regional consciousness. More intensive studies at the local level could reveal how such factors operate.

The studies just noted were based on insiders' views of the region in which they live. Some of the discussion that follows examines other aspects of perceived regional structure, such as how people perceive regions other than the one they live in.

Student schemata of the United States

Kevin Cox and Georgia Zannaras developed a methodology for investigating the sets or subclasses students use in their mental regional organization of the United States.[6] A small group of University of Ohio students were asked to take each state and select the three states most similar to it. Later, by using the technique of factor-analysis, the results were evaluated to see the manner in which the students grouped the various states.

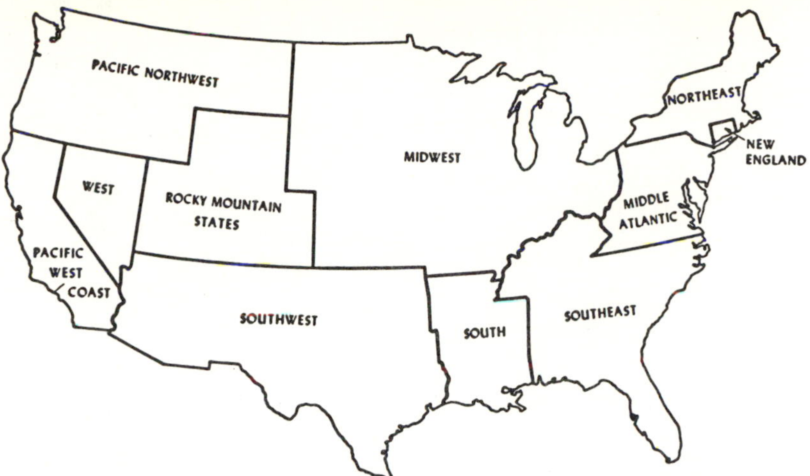

Fig. 7.2 Vernacular national regions. (Ruth F. Hale, "A Map of Vernacular Regions in America," unpublished dissertation, 1971, p. 66.)

The strongest differentiating feature appeared to separate the states on an East-West basis, with the Mississippi River serving as a boundary line. A second factor differentiated between the South and New England with a Southern Borderland and a "Deep South" distinguishable. In addition, there was a perceived area of homogeneity in the Midwest, a Midcontinent region, an Appalachian factor, and a contrast between the Pacific Northwest and the Southwest (Figure 7.4).

Cox and Zannaras suggest that

> . . . it is possible to conceive of a system of classes or regions corresponding to the schemata of the students by placing the areas of homogeneity revealed by the factors into some sort of simple hierarchical classification: on the one hand we have the West which can be broken down into the Pacific Northwest, the Mountains, the Southwest, and the Midcontinent; on the other hand we have the East comprising the South, Appalachia, the Midwest, and New England.[7]

The Cox and Zannaras study, though focusing at the national level, is basically concerned with investigation of the general properties of mental maps discussed in Chapter 5. In particular, they have tried to understand the criteria people use in classifying places. One criterion investigated more fully than most of the others is that of preference, discussed in the following section.

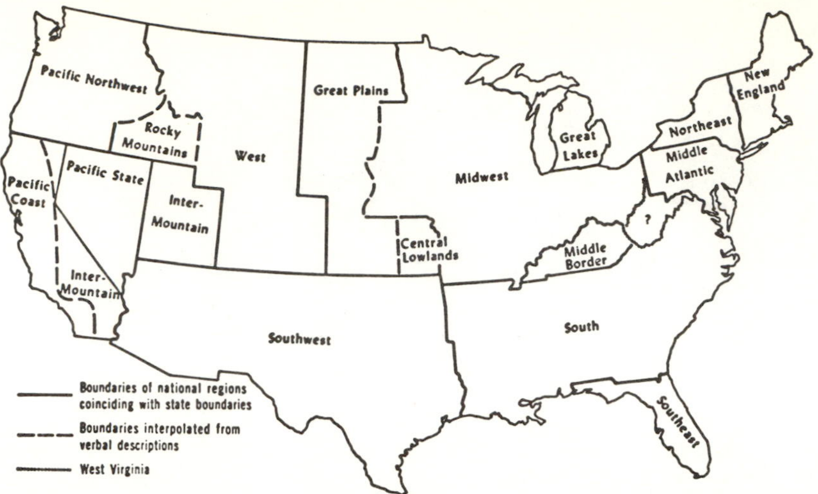

Fig. 7.3 Geographers' national regions. (Ruth F. Hale, "A Map of Vernacular Regions in America," unpublished dissertation, 1971, p. 118.)

Studies of place preferences

Peter Gould, a geographer at the Pennsylvania State University, has stimulated a line of research focusing on people's place preferences. His original study utilized samples of students from a number of American universities.[8] They were asked to list in order their preferences for states in the United States. The question was posed in terms of residential desirability. Based on the data on preferences, isoline maps were constructed to represent the relative desirability of various areas to the students. A remarkably similar map of the United States emerged for most groups of American students. The West Coast is seen as the most desirable area. From a high here, the surface slopes downward to the Utah perceptual basin, rising once more to the Colorado high. Over the Great Plains there is a general decline eastward, with a low point in the Dakotas. Near the one-hundredth meridian a change occurs, with a rise toward the northeast and a drop in desirability toward the South, the lowest perceptual trough of the entire surface. The map of the California students corresponds well to common elements (Figure 7.5) found on the preference maps from Minnesota and Pennsylvania. The only major exceptions are that home areas receive the highest ratings. However, the maps of Alabama students show a different set of preferences (Figure 7.6). From their perspective the North is seen as undesirable; the South is more highly differentiated, with Alabama the most desirable; and a new low appears over New Mexico. Clearly, the

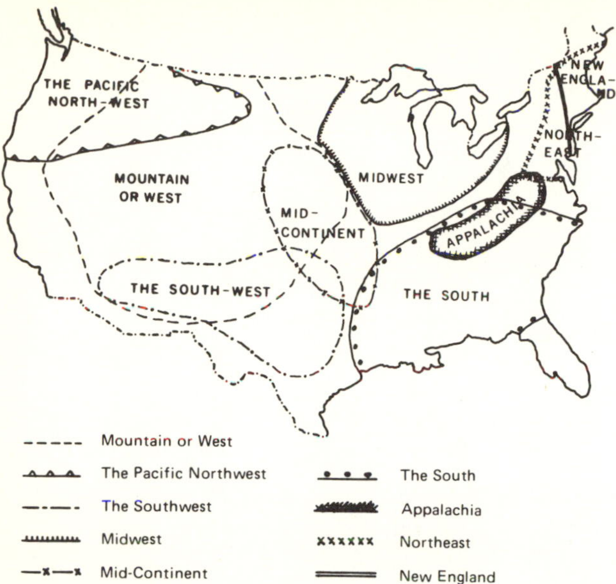

Fig. 7.4. Locational classes of student schemata of the U.S. (Kevin R. Cox and Georgia Zannaras, "Designative Perceptions of Macro-Spaces" in *Image and Environment*, eds. Roger M. Downs and David Stea, Aldine Publishing Co., Chicago, 1973, p. 171. Reprinted with permission.)

home area ranks high in such preference maps. But Brian R. Goodey's similar study in North Dakota indicates that there the home state is not seen as the most desirable.[9] However, he found that the most familiar states, that is, the ones most often visited, showed up as the most desirable, and these were also the ones to which North Dakotans most generally migrated. Peter Gould and Rodney R. White applied the same sort of procedure to determine British school-leavers' residential preferences[10] and J. M. Doherty did a similar study of preferences for the cities of the United States.[11] Later studies have examined the pattern of preferences in Tanzania,[12] the Western Region of Nigeria, and Sweden.[13] The focus has shifted to a consideration of the development of geographic space preferences through different age groups and on the flows of information from which such images are formed. Preference maps may help explain migrations and movements of people within a country or, if prepared on other scales, world, city, or regional migrations.

Location preferences and migration

Niles Hansen in *Location Preferences, Migration, and Regional Growth* shows some practical implications of the study of people's location preferences.[14]

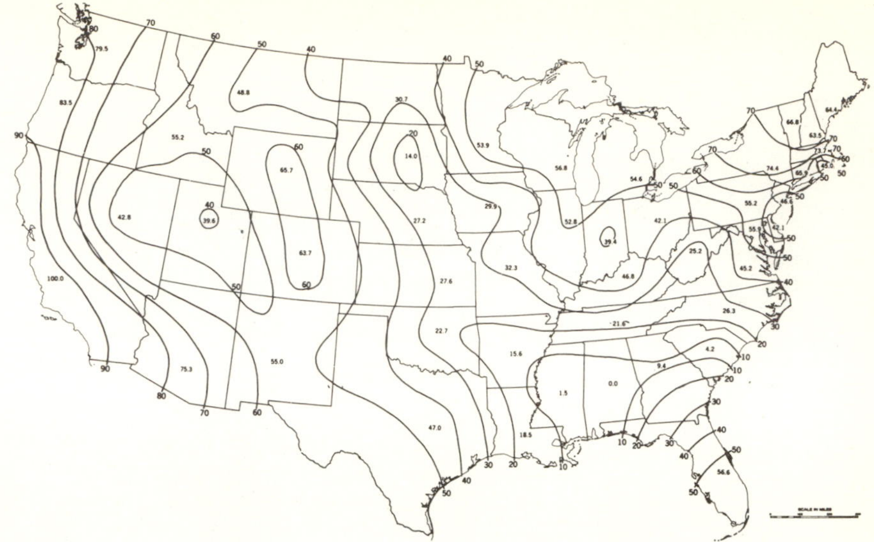

Fig 7.5 The view from California. (Ronald Abler, John S. Adams, and Peter Gould, *Spatial Organizations: The Geographer's View of the World,* © 1971. Reprinted by permission of Prentice-Hall, Inc., Englewood Cliffs, New Jersey.)

He states, "it is remarkable that so little attention has been given to the residential location preferences of individuals or how the geographic organization of economic activities may be brought more into line with the preferences of people."[15] On the assumption that people wish to remain where they are, there has been a general policy to promote economic activity in lagging areas. Hansen argues that it might be more efficient to give more emphasis to human resource development and promotion of labor mobility. But he feels that migration to the already congested large cities should be discouraged. Instead, he favors a program of development assistance to intermediate-sized cities closer to the lagging areas.

> . . . if a federal subsidy can accelerate growth in a center that is already growing, and if this subsidy is made conditional on providing employment opportunities for the residents of lagging areas, then it could well be more efficient to tie into the growing environment than to attempt to create growth in a relatively stagnant area. Of course the feasibility of this approach depends in part on the location preferences of people in lagging areas. Are there a significant number of people in these places who want to move? If so, under what conditions and to what kinds of places do they want to move?[16]

To investigate the questions raised, Hansen conducted surveys in a number of poor rural areas of the United States. The people surveyed were mainly high school seniors from eastern Kentucky, south Texas, southwestern Indian reservations, and southwest Mississippi. This age group

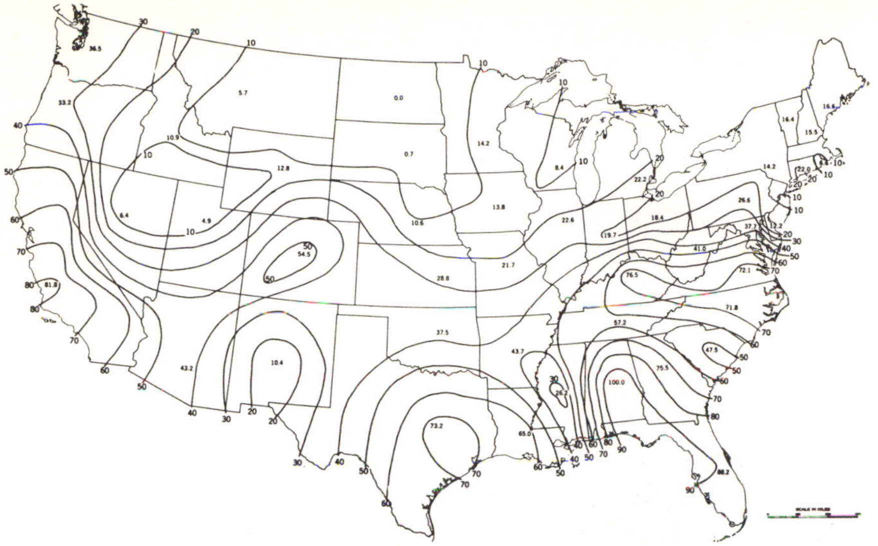

Fig. 7.6 The view from Alabama. (Ronald Abler, John S. Adams, and Peter Gould, *Spatial Organizations: The Geographer's view of the World,* © 1971. Reprinted by permission of Prentice-Hall, Inc., Englewood Cliffs, New Jersey.)

was selected because it seemed likely to have a relatively high mobility potential. The frequency of preferences for home areas, intermediate centers, and large cities was established for various different combinations of wages. Table 7.2 summarizes the results.

The majority of people appear to prefer the home area, given the possibility of equal wages in each place. However, even under these conditions a substantial proportion would prefer to migrate elsewhere. The preference patterns are definitely influenced by variations in wages. This is apparent in each case where the wages are better in one type of location. By comparing Case VI with Case VII, one can see clearly that the intermediate centers are preferred to the large cities. In these cases the wages in the home area are held constant, while those in the intermediate and large cities are reversed. Even where there is a clear advantage in wages in the large city, large proportions of those willing to move tend to prefer the intermediate centers. Hansen's data do indicate that there is some potential for the type of regional development policy he favors, but they also underline the strength of preferences for the home area and for nearby cities.

For most centers it seems likely that the greatest potential source of new migrants is the immediately surrounding region. This seems to be a natural outcome of the tendency to prefer the familiar and places close to home. From a national point of view it may also be the wisest, as Hansen suggests, for a shorter move may mean a less difficult adjustment because of the

Table 7.2 Relative Frequency of Preferences for Home Areas, Intermediate Centers, and Large Cities by Selected Groups of Young Persons

Case and location preference		Hourly wage ($)	E. Ky. (1969) (%)	E. Ky. (1971) (%)	Mexican Americans in S. Tex. (%)	Southwestern Indians (%)	S.W. Miss. Blacks (%)	S.W. Miss. Whites (%)
I	Home area	1.50	64	69	75	43	65	74
	Intermediate center	1.50	26	24	15	29	15	21
	Large city	1.50	10	7	10	28	20	5
II	Home area	1.50	25	24	42	19	33	45
	Intermediate center	1.75	36	38	23	38	22	40
	Large city	2.00	39	37	35	43	45	15
III	Home area	1.50	18	16	31	17	20	28
	Intermediate center	2.00	40	40	34	38	30	54
	Large city	2.50	42	44	35	45	49	18
IV	Home area	1.50	10	14	21	16	16	18
	Intermediate center	2.25	36	37	38	33	27	56
	Large city	3.00	54	50	41	51	58	26
V	Home area	3.50	81	84	84	61	82	83
	Intermediate center	2.50	13	14	11	25	9	14
	Large city	1.50	5	3	5	14	10	3
VI	Home area	1.50	10	10	20	16	12	7
	Intermediate center	3.50	81	86	76	64	74	88
	Large city	2.50	9	4	4	20	14	5
VII	Home area	1.50	9	12	20	13	14	14
	Intermediate center	2.50	35	36	37	35	27	56
	Large city	3.50	56	52	43	52	59	30
VIII	Home area	1.50	6	9	14	15	8	8
	Intermediate center	3.50	27	27	30	32	18	45
	Large city	5.50	66	64	56	52	74	48

SOURCE: Niles Hansen, *Location References, Migration, and Regional Growth*, © 1973 by Praeger Publishers, Inc., New York, p. 132. Reprinted by permission.

greater similarity of society in adjacent areas and the possibility of retaining home ties. Although the majority of migrants may prefer areas close to home, certain of the nation's regions may stand out as pre-eminently attractive. Such areas may draw migrants, perhaps of a different sort, from the entire nation. A good example is California, long the mecca of internal migration in the United States.

Geography of the ideal

James E. Vance, Jr., interprets the settlement of California in terms of the geography of the ideal.[17] He argues that internal migration in America has always fed upon images, and nowhere is this more clearly illustrated than in the case of California, which was settled largely by those searching for

the desirable life style. The very name *California* was coined to represent an imaginary island very near the quarter of the terrestrial paradise. The favorable image was enhanced with the Gold Rush of 1849, and, when this ended, new images were created, billing California as the land of restorative climate and as an agricultural arcadia, and later as an area foreign to American norms and tolerant of social peculiarities. The pull of the ideal could best be afforded by the well-to-do, and these were the dominant group in the first wave of American migration to Southern California. They were followed in each succeeding wave by people of progressively lesser means and status. Vance states that since 1946

> . . . southern California has largely lost its image as the land of the ideal, and subsequent migration thence has come to reflect a more normative pull, that of money, whereas the geography of the more illusive image has shifted successively poleward, first to the Bay Area (1945-1960), then to rural northern California (1960-1970), and, it seems now, northward to Oregon, Washington and British Columbia (1970-   ).[18]

By becoming aware of the geography of the ideal, one may gain a better understanding of America today and of trends likely to appear in the future in other areas of the world. The types of migrants attracted by the images of California would probably differ in many respects from those likely to move to areas closer to home. Much remains to be learned about which types of people would move to what kinds of areas and for what reasons. The answers to such questions might reveal a good deal about the types of landscape that would be likely to evolve in various places.

Landscapes and landscape tastes

David Lowenthal and Hugh Prince collaborated in a pair of early articles that linked the appearance of national landscapes to people's preferences.[19] Their thesis, as seen in the opening quotation of this chapter, is that "landscapes are formed by landscape tastes." What people perceive as the ideal is what they aim at as they transform the landscape. Lowenthal and Prince sought the idealized images of the past and present by examining travelers' accounts, landscape paintings, literature, and attitudes expressed in speeches, public hearings, newspaper articles, and letters. They assumed that the articulate minority responsible for such source materials are the people most influential in creating landscape tastes and hence in molding the landscape.

According to Lowenthal and Prince, the English taste in landscape may be seen by examining the English landscape, which reflects the English preference for the bucolic, the picturesque, the deciduous, the tidy, façadism, antiquarianism, and the uniqueness of each place. The seem-

ingly natural landscape of rural England has not occurred by chance. Rather, it reflects the English taste for the deciduous, the bucolic, and the picturesque. The trees are planted where they are to achieve the effect desired. The love of a good ruin combines the taste for the picturesque, for façadism and antiquarianism. During certain periods deliberate destruction to create such ruins was not uncommon. Concern for façadism has led to the dressing up of the landscape in various ways to create the proper appearance. In a later article Lowenthal and Prince document many examples of façades contrived to make landscapes look attractive, appropriate, or inoffensive.[20] For example, the Forestry Commission screens plantations of conifers by planting fringes of deciduous trees, or street façades maintain period treatment though interiors belie the outward appearance.

Lowenthal and Prince suggest that the key to summarizing the relationships between the English people and their landscape is the word *amenity*. This term is attached to whatever seems to need protection. Thus amenities may at various times be historic buildings and the flavor of the past, open spaces, views and vistas, facilities for recreation, or access to points of scenic interest. Government departments and private interests exhibit the English talent for compromise, and the result is a settled and comfortable land.

In contrast to the settled and comfortable English landscape is the American scene, also described by Lowenthal.[21] He states that visitors generally see the American landscape as "vast, wild, and empty, formless and unfinished, and subject to violent extremes."[22] Other writers have also produced well-illustrated polemical books describing visual blight in America.[23] But most Americans probably see the landscape in other ways. Those who settled the land saw it not as scenery but as something to be tamed and made useful. The landscape was scarcely seen at all, but only acted upon. The important thing was that it served its purpose in producing crops or other products. Even today Americans tend to see their landscape not as it is but as it is going to be. The present is sacrificed to some vision of future greatness. Scenic appreciation is highly serious and self-conscious, but reserved for the remote and spectacular. Meanwhile, the nearby and familiar is neglected. Individual features are emphasized while the total setting for them is ignored. This produces the disconnected appearance commented upon by visitors.

The Lowenthal and Prince thesis that landscapes are formed by landscape tastes underlines the essentially subjective nature of beauty or ugliness. A recent paper explores the kind of understanding geographers must gain to launch an effective action campaign against the growing problem of visual blight in America.[24] Pierce Lewis states:

> Aesthetic judgement is inherently perilous, and most scholars know it. In the great Platonic triad of virtues—Truth, Goodness, and Beauty—scholars grow increasingly nervous as they pass from one to another. The search for Truth, of course,

is the scientist's stock-in-trade. As for Goodness, although Jeremy Bentham had trouble measuring it, most scholars uncritically accept a variety of moral strictures without undue worry. But judgements about Beauty seem quite another matter. Immeasurable, slippery, and quintessentially subjective, the whole area of aesthetic judgement has traditionally been left to art critics, and studiously avoided by geographers and most other social scientists.[25]

Evaluations of landscapes based on sound aesthetic standards are difficult and subject to change. In addition, the problem of subjectivity in landscape taste is further complicated because standards other than aesthetic ones are commonly applied. Yi-Fu Tuan points out that aesthetic values, including concepts of spatial harmony, as in landscapes, are things we acquire in the process of maturation and learning.[26] Ultimately, it is the great artists who teach us as they accustom our eyes to see the environment in new ways. But there have been tides of taste among the great artists as well. The changes over time in the way artists have viewed landscapes is well illustrated in Kenneth Clark's book, *Landscape Into Art.*[27] The standards set by the great artists, however valuable, are not the sole basis of landscape evaluations. Aesthetic judgments of landscapes are often permeated by social, moral, and ecological values. People may become attached to a particular physical environment because they have labored over it to gain a livelihood or because it exemplifies an ecological ideal.

> By "visual blight," we may simply mean that we are judging the health of society on the ground of visual evidence in the landscape. We see smashed beer bottles on the sidewalk and dark tenements with broken windows, and take them as evidence that society is desperately ill. We see broad expanses of well-tended lawn and discern that they offend against the ecological principle of species complexity. In other words, the argument as to what makes for quality in an environment is pursued at the level of social and moral philosophy and of ecological principles, and not at the level of aesthetics.[28]

Harmony between people and their environment may be reflected in a beautiful landscape. But creating such a landscape is not simple. It involves a heightened sensitivity to place and depends on many individual decisions. Donald Meinig, speaking to geographers, argues that:

> . . . if we really want to deal with Visual Blight as serious scholars as well as political activists we shall have to delve very deeply into such elusive topics as "a sense of place," the "meaning of landscapes," "environmental influence," the concept of "home," the feelings, the moods, the deep emotional reactions to the particular character of particular places.[29]

For geographers such studies would involve a more thorough immersion in the humanities. For those in the humanities the same end could perhaps be achieved by gaining a greater understanding of the geographic perspective on places.

Every national landscape might be expected to reflect the attitudes and ideas of the citizens or government. This theme has been considered in a number of recent monographs on different national landscapes. An example is W. H. Parker's discussion of the changing Soviet attitudes toward the environment.[30] He notes a number of stages. In the first years of Soviet power, Lenin's Marxist views prevailed. All regions of the country were to be developed according to their natural conditions, with industry located close to energy sources and raw materials. But during the 1930s there was a change as excessive bureaucratic centralization in Moscow led to a growing disregard for local and natural conditions. This in turn was succeeded by a shift back to excessive regional self-sufficiency as a result of the threat of war in the late 1930s and a continuance of the policy through the 1940s and 1950s.

In the period of Stalin's rule, the prevailing attitude was "voluntarism," implying that the Soviet society could do as it pleased in defiance of nature. Stalin viewed the geographic environment as the essentially static nature that surrounds dynamic society. Ambitious and costly schemes linked with personal glorification of Stalin were pushed through. Among other "Great Stalinist Plans for the Transformation of Nature" were planting tree belts in the steppes, converting the Volga into a series of lakes linked by hydroelectric dams, irrigating vast areas of arid land, changing the channels of the Ob and the Yenisei to water the deserts of Central Asia, establishing nuclear power stations to heat the waters of the Arctic, and schemes to change the climate of Siberia. Since Stalin's death in 1953, voluntarism has been officially and academically discredited—although Khrushchev was also criticized for "wild and harebrained schemes"—especially for his continuation of the Stalinist attitude in his agricultural policy. Now there is a greater emphasis on local studies before making decisions on land use. This may have been influenced by the unforeseen consequences of changing the landscape. An example of the new approach would be weighing the loss of farmland and damage to the Caspian fishing industry against the increased power and irrigation of Volga development plans. Grandiose schemes are still being advanced, but today there is greater discussion of the pros and cons. Like people in other advanced industrial nations all over the world, the Soviet people may be forced to move toward policies more in harmony with nature. Whether this may be easier in a centrally planned economy than in a free enterprise system seems to be a matter of some dispute.[31]

The American environmental movement

The shifts in Soviet attitudes and policies challenge the assumption that attitudes toward the environment are fixed and immutable. The rapid

emergence of ecology as a national issue within the United States also belies such an assumption. Hazel Erskine, polls editor of *Public Opinion Quarterly*, said in 1972:

> A miracle of public opinion has been the unprecedented speed and urgency with which ecological issues have burst into American consciousness. Alarm about the environment sprang from nowhere to major proportions in a few short years. When the first polls on pollution appeared in 1965, only about one in ten considered the problem very serious. Today most people have come to that realization.[32]

Table 7.3 reviews the results of several polls asking the same question in later years. There is no doubt that a spectacular change in public opinion has been charted. This holds true whether one considers the total national sample or the perspectives of residents of big cities or major geographic regions. The growth in awareness is paralleled by an increase in the proportion of people who express a willingness to pay for pollution control.

Increased awareness and concern are not limited to the problems of air and water pollution. Membership growth rates in conservation and environmental preservation organizations of all types have shifted sharply upward. The Sierra Club, for example, had a membership roster of about 7,000 in 1952. But by 1970 the number of members had skyrocketed to over 100,000.[33] Discussion of problems of the urban environment, of outdoor recreation, and of threatened animal species appeared with increasing frequency in newspapers and magazines and on radio and television. The discussion has extended to all the problems, present and anticipated, that result from rapid growth of population, urbanization, industrialization, and affluence.

The rapidity of this apparent change in public opinion has stimulated a great deal of discussion. Some researchers have tried to determine which social groups support environmental issues. Other studies have tried to explain in terms of the societal system how the change took place and why it emerged with such strength in the United States as opposed to other world areas. Some observers express doubt as to the depth or sincerity of the professed environmental concern. They predict problems that are likely to arise as the full implications of improving the environment sink in.

The 1969 poll by the Gallup Organization for the National Wildlife Foundation provides a measure of which Americans are concerned about their environment.[34] Table 7.4 illustrates the answers to the question:

> You may have heard or read claims that our natural surroundings are being spoiled by air pollution, soil erosion, destruction of wildlife and so forth. How concerned are you about this—deeply concerned, somewhat concerned, or not concerned?

The strongest single factor for predicting environmental concern would appear to be educational level. The more highly educated express greater

Table 7.3 Concern Over Pollution

Compared to other parts of the country, how serious, in your opinion, do you think the problem of air/water pollution is in this area—very serious, somewhat serious, or not very serious?

	Very, Somewhat Serious	
	Air	Water
1965: May	28%	35%
1966: November	48	49
1967: November	53	52
1968: November	55	58
1970: June	69	74
Big city residents		
1965	52	45
1966	70	59
1967	76	62
1968	84	73
1970	93	89
	Very Serious	
	Air	**Water**
By geographic region:		
Northeast		
1965	20%	21%
1967	29	30
1968	34	35
1970	51	53
Midwest		
1965	8	14
1967	29	28
1968	26	35
1970	33	41
South		
1965	3	9
1967	14	18
1968	12	18
1970	20	27
West		
1965	13	6
1967	42	17
1968	37	22
1970	42	28

SOURCE: Hazel Erskine, "The Polls: Pollution and its Costs," *Public Opinion Quarterly*, 36 (Spring, 1972), No. 1, 120-135, table on p. 121. Reprinted by permission.

environmental concern. The less educated and those of lower socioeconomic status have the highest proportion who are not very concerned. This lower level of concern among the less affluent may result from their focus on other more pressing problems they must deal with each day. A parallel situation exists on the world scale—representatives of the more

affluent nations are clamoring for pollution controls while the poorer developing nations place other issues higher in priority.

The sociologists Denton E. Morrison, Kenneth E. Hornbeck, and W. Keith Warner explain the rise of the environmental movement in the United States in terms of the social-psychological theory of "relative deprivation."[35] Social movements generally emerge and receive their early support from those who have come to expect a substantial measure of the movement's goals. When, rather suddenly and unexpectedly, blockages appear in the existing conventional routes to the goals, these people develop the degree and kind of social discontent that predisposes them to join with others in an urgent attempt to bring about change. Thus the United

Table 7.4 Distribution of Responses to a National Survey of Public Concern over Environmental Problems (Percentage)

	Level of environmental concern					
	Deeply concerned	**Somewhat concerned**	**Not very concerned**	**No opinion**	**Total**	**Number of interviews**
National Results	51	35	13	2	100	1503
By Sex						
Men	56	31	10	3	100	744
Women	46	38	14	2	100	759
By Age						
21-34 years	51	41	7	1	100	403
35-49 years	50	38	10	2	100	476
50-years and older	52	28	16	4	100	605
(Undesignated-19)						
By Education						
College	62	32	6	*	100	395
High school	52	37	10	1	100	748
Grade school	39	34	20	7	100	352
(Undesignated-8)						
By Annual Family Income						
$10,000 and over	58	34	8	0	100	449
$7,000-$9,999	53	38	8	1	100	336
$5,000-$6,999	55	35	8	2	100	237
Under $5,000	41	34	20	5	100	463
(Undesignated-18)						
By Size of Community						
1 million and over	51	36	8	5	100	277
250,000-999,999	52	35	11	2	100	296
50,000-249,000	55	35	9	1	100	235
2,500-49,999	52	31	16	1	100	233
Under 2,500	46	37	14	3	100	462
By Region of Country						
East	46	38	2	4	100	425
Midwest	56	34	9	1	100	400
South	44	36	16	4	100	428
West	59	31	10	*	100	250

*Less than half of one percent. SOURCE: The Gallup Organization, Inc., "The U.S. Considers its Environment," *National Wildlife Magazine*, February, 1969, p. 7. Reprinted by permission.

States, the prime locus of the environmental crisis, is a nation where people have access to better rather than poorer environments; and, as we have just noted, it is the more affluent within nations who are most concerned. The crisis perceived by the environmentalists is relative to their higher expectations for their environment.

> The environmentalists have largely solved certain basic economic problems and have both the leisure and the conceptual tools (through their formal education and their extent and kind of media exposure) to take a broader, longer-range view of the environment to gain the perceptions of crisis that such a view implies.[36]

Resistance to costly environmental reforms may be expected from those who see their own economic expectations blocked by such reforms. Labor, management, and owners of threatened industries are organized groups that might offer such resistance. The analysis of the environmental movement by Morrison *et al.* is useful in predicting likely developments that could help or hinder future efforts to improve the quality of our environment. By foreseeing trends in social behavior, more realistic policies and plans may be adopted that go beyond acceptance at face value of opinion polls.

Several observers have expressed some doubt as to the depth of interest in and degree of commitment to environmental issues. Lowenthal, for example, questions whether even the most vocal environmental advocates appreciate what real reform might entail.[37] He argues that stated values about landscape and nature often run counter to behavior. Lowenthal feels that landscapes reflect social values that are taken for granted. He doubts that such values will be sacrificed for the sake of environment. Certainly the government commitment in terms of dollars invested for environmental cleanup falls far short of the fervor of the rhetoric. W. R. Derrick Sewell and Harold D. Foster have also summarized many of the contrasts between idealism and action.[38] They suggest that a better understanding of the discrepancy between concern and commitment may help us predict individual response. Although there is greater general awareness of the problem, the same social and economic factors that created the crisis remain and may prevent its solution. Major changes in habit and organization are required, and these will be more difficult to achieve than the technological knowledge necessary to solve pollution problems.

The fundamental conflict between the aims of the production system and those of environmental quality has often been noted. It is clear that the very processes that have created the material goods associated with affluence have also enlarged the load of wastes that must be absorbed by the environment. What we have is a gigantic garbage disposal problem aggravated by increasing numbers of people. As the level of affluence rises, each person has increasing amounts of garbage to dispose of. To end the crisis, we must change our consumption habits, and very few observers consider this an easy task. Most of the major American corporations are based on

energy production or automotive-type industries. Hence an enormous struggle would be necessary to shift to an ever less energy-profligate industrial society, even if most of the public were willing to change their own consumption patterns.

National character and environment

Many of the doubts regarding the depth of change noted in connection with the environmental movement in the United States may be attributed, at least in part, to a belief in the enduring qualities of national character.[39] In spite of variations by age, sex, social class, and other group and regional differences, most people within a nation share a common set of traits that set them apart from other nations. This national social character transcends group and individual differences. It shows up in characteristic attitudes and evaluations, common conceptions or misconceptions of the natural and social world, and notions of appropriate and inappropriate ways of solving life's problems. National character is normally not noticed in day-to-day activities within one's own nation. But it becomes immediately noticeable as one moves to another nation and has dealings with local people there. On a world level it is exceedingly important to understand national character.

Some of the international effects of national character are explored further in Chapter 8. Here the significance of national character will be briefly discussed in terms of how the distinctive American ethos is reflected in the landscape and the environmental behavior in the United States.

In a recent book, *The Cultural Geography of the United States,* Wilbur Zelinsky isolates four pervasive themes or motifs that reflect the geographically relevant, locally distinctive character of American culture.[40] These are: (1) an intense, almost anarchistic individualism; (2) a high value placed on mobility and change; (3) a mechanistic vision of the world; and (4) a messianic perfectionism.

> The heroic self-image of the lone, self-reliant, upward-striving, individual, sharing equal rights and opportunities with all and liberated from the shackles of tradition and authority, is possibly the single most dominant value in the cultural cosmos of the American. This fanatical worship of extreme individualism, indeed an almost anarchistic privatism, affects so many phases of our existence so deeply that no one can interpret either the geography or the history of the nation sensibly without coming to grips with it.[41]

The American stress on individualism shows up in settlement patterns, in economic patterns, in religion, and in political behavior. The predominant and idealized rural settlement pattern is one of isolated individual farms with the self-sufficient farm family. This may underlie much of the

American distrust of the city and the nostalgia for small-town and rural life discussed in previous chapters. The multitude of separate religious denominations is unparalleled anywhere else. The supremacy of the private car is unquestioned. The total reliance on the free enterprise system shows up in the lack of any effective national railroad system or national airline, a rare situation for a large modern nation. With the extreme individualism comes a fear and suspicion of authority, so that many of the functions most efficiently performed by a national government are instead handled by a fragmented set of state, county, and other local organizations. Many of the problems noted in previous chapters may be attributed to the general American distrust of long-range social and economic planning of all kinds.

The high value placed on mobility and change may be seen in the innate restlessness of Americans. Their spatial mobility appears not only in constant change of residence and long-range migration but in daily circulation patterns. The emphasis on haste, speed, and change is associated with a lack of appreciation of past landscapes, which are blithely bulldozed away to make way for the current controlling element, the highway. "Cruising for pleasure" along the highway has led to the highway strip developments discussed in Chapter 4, with their characteristically American inventions, the drive-in establishments.

The mechanistic world vision of Americans assumes that the universe is a fundamentally simple, mechanical system subject to human control, a machine that can be improved and accelerated.

> Within the adjustable world-machine envisioned by Americans, the human actor is a detachable, freely mobile, perfectible cog; . . . the interchangeable Yankee, highly versatile and moveable, and eager to insert himself into the locus of maximum advantage to himself and thus, of course, to the system. The phenomenal social, economic, and spatial mobility of the individual thus becomes a national way for lubricating and putting into a state of frantic productivity all those gears, wheels, levers, springs and pulleys that comprise the physical and social world.[42]

A mechanistic world is a quantifiable one, and Americans eagerly count and measure all aspects of their universe. Their land and cities are divided and numbered by the rectangular land survey and the grid system of streets. This strong emphasis on numbers is basic to the beliefs that "growth is good" and "bigger is better." Skyscrapers are symbols of this cult, as are big dams and projects. The inadvertent side effects of large environmental interventions indicate the simplistic nature of the mechanistic notion. But even as these side effects appear, new faith springs up in the ability of technology to solve each new problem.

> Perhaps the most extraordinary facet of the American national character is the power and durability of an idea only weakly, fitfully developed in other cultures: the notion that the United States is not just another nation, but one with a special

mission—to realize the dream of human self-perfection and, in messianic fashion, to share its gospel and achievement with the remainder of the world.[43]

This messianic streak is apparent in American foreign policy and in such American phenomena as the Peace Corps. It has been expressed in the landscape in such place names as New Hope, New Jerusalem, and Zion, as well as in the scores of utopian communities established in America from the earliest New England settlements to the present. The geography of the ideal discussed earlier in this chapter provides another example.

National character offers a key to understanding human behavior in a nation. In the case of the United States, many of the characteristic social traits, ideas, and attitudes have a clear expression in the landscape and environmental behavior of Americans. But there are major problems involved in defining national character, and how it is developed, and how to measure it. There is great difficulty in comparing systematically the large numbers of people who comprise a nation. In spite of these difficulties, it is hard to deny that there is such a thing as national character or that it is of fundamental importance. On a world level the importance of understanding national character becomes even more marked. In the preface to a monograph on national character in the perspective of the social sciences, Don Martindale expressed the importance as follows:

> The more fully they understand their national social characters the more possible it is for contemporary men also to overcome their unique parochialism and to devote their human and physical resources to the formation of a world community to benefit all mankind.[44]

The parochialism he speaks of is demonstrated in several studies described in the following chapter.

Summary

The segment of environment considered in this chapter is immense, far beyond the scale one can see and comprehend at a glance. The studies reviewed focus on people's conceptions of a complex reality and on generalizations about the behavior of very large groups of people. The large numbers of people considered and the complexity of the environment require new methods of of measurement. Thus we see the use of public opinion polls to measure people's attitudes toward environment. A greater reliance is also placed on the use of the kinds of insights and generalizations derived from the humanities. The views of novelists, artists, journalists, poets, and historians are canvassed in exploring national attitudes and their interaction with environment. Such views often contain compelling images or ideas that symbolize the situation at the level of abstraction required. The importance of symbols and images at the scale of the nation

may be seen in discussion of such topics as the geography of the ideal and the types of landscapes valued by the English and the Americans. The greater degree of generalization and abstraction that leads to concern with symbolism and images is a characteristic shared with many studies at the scale of the city and larger units. It is not quite so common in research on the smaller-sized segments of the environment, where direct perception and behavior can more easily be measured.

The environment as conceived at the national level differs somewhat from that at other scales. The natural and built environments tended to be merged in such concepts as national landscapes and attitudes toward the environment. The interdependence and connectedness of the artificial and natural environments show up clearly as a total system. This could be considered a national ecosystem with complex interactions among the various elements. The social environment is also much more abstract than in earlier chapters. Here, the concern is with large social groups, distinguished on the basis of age, social class, and geographic region. Their varied perspectives and images of the environment may be described or analyzed with respect to their position within the national social structure.

In spite of the differences due to the change of scale, several familiar approaches and problems appeared once more. When considering national policy for regional development, Hansen noted a phenomenon remarked on by researchers at all scales. This is the familiar pattern of planning without consulting those most affected as to their needs and preferences. Assumptions have been made about where people might like to live without resort to the obvious measure of asking them. As in previous chapters, several studies were devoted to investigating the characteristics of cognitive maps. Squarely within this category would be the student schemata of the United States studied by Cox and Zannaras and Gould's work on student place preferences. Highway strip developments, noted first in Chapter 4, may be placed in a broader national perspective as one facet of this chapter's concern with the American landscape and landscape tastes. The concept of regional consciousness was developed more fully here, but the examples provided indicate that the same set of feelings may operate at several different scales. At the same time an individual may feel a strong sense of belonging to a particular neighborhood and to the nation.

Examples of different professional perspectives appeared once more. The familiar focus of geographers on landscapes was represented by several authors. Morrison and his associates reflected the sociologist's research on social groups and their interactions. The psychological perspective, emphasizing the individual, and the anthropological concern with the role of culture were both present. However, at the national scale they were merged in the topic of national character, where studies of personality and culture overlap.

The broad, pervasive influence of culture has been referred to repeatedly in each chapter so far. Its importance cannot be overemphasized. At the

national level exceedingly powerful forces work to reinforce the cultural values of the group. The results are reflected in the broad general characteristics of people referred to as national character, and in the frequency with which certain shared attitudes tend to be present. Such culturally conditioned behavior as the use of space described by Hall in Chapter 2 could be interpreted as observable aspects of national character.

In this chapter national character was discussed in terms of certain recurrent themes commonly used in descriptions of Americans and their behavior. The emphasis was on environmental attitudes and behavior and their reflection in the American landscape. As in much of the work on national character, the discussion proceeded from judgments of what constitutes the character to illustrations that served to corroborate the judgments. The degree to which the judgments reflect reality will depend on the insight and understanding of the person making them. To improve the chances of success in such an enterprise, the inside, subjective knowledge of the culture must somehow be combined with the often more objective perspective of an outside observer. At best, such descriptions can provide profound insight into the character of a country. At worst, they degenerate into national stereotypes of a negative, derogatory nature. A major problem in many conflict situations today is that such stereotypes, simplistic and silly as they often seem, may be acted on as if they were true. This topic is developed more fully in the following chapter, as we step up in scale to the international level.

Notes

1. A fine discussion of regional consciousness is found in Robert E. Dickinson, *Regional Ecology: The Study of Man's Environment,* John Wiley & Sons, Inc., New York, 1970, pp. 54-58.
2. For a discussion of names that indicate the role of the perceptions of the namers, see George R. Stewart, *Names On the Land,* Houghton Mifflin Company, Boston, 1967.
3. Mauri Palomaki, "On the Concept and Delimitation of the Present-day Provinces of Finland," *Acta Geographica,* 20 (1968), 279-295.
4. Ruth F. Hale, "A Map of Vernacular Regions in America," unpublished Ph.D. thesis, University of Minnesota, June, 1971.
5. Carl F. Kraenzel, *The Great Plains in Transition,* University of Oklahoma Press, Norman, Okla., 1955, pp. 212-226.
6. Kevin Cox and Georgia Zannaras, "Designative Perceptions of Macro-Spaces: Concepts, A Methodology and Applications," *EDRA Two,* Proceedings of the Second Annual Environmental Design Research Association Conference, October, 1970, pp. 118-130.
7. *Ibid.,* p. 125.
8. Peter Gould, *On Mental Maps,* Michigan Inter-University Community of Mathematical Geographers Discussion Paper No. 9, Department of Geography, University of Michigan, Ann Arbor, Mich., 1966.
9. Brian R. Goodey, "A Pilot Study of the Geographical Perception of North Dakota

Students," unpublished paper, University of North Dakota, Department of Geography, 1968.

10. Peter Gould and Rodney R. White, "The Mental Maps of British School Leavers," *Regional Studies,* 2 (1968), 161-182.
11. J. M. Doherty, *Residential Preferences For Urban Environments in the United States,* London School of Economics Graduate School of Geography Discussion Paper No. 29, London School of Economics, London, 1968.
12. Peter Gould, "The Structure of Space Preferences in Tanzania," *Area,* No. 4 (1969), 29-35.
13. Peter Gould, "Geographic Exposition Information and Location," in *National Academy of Sciences Geographical Perspectives and Urban Problems,* National Academy of Sciences, Washington, D.C., 1973 pp. 25-40.
14. Niles Hansen, *Location Preferences, Migration, and Regional Growth,* Praeger Publishers, New York, 1973.
15. *Ibid.,* p. 65.
16. *Ibid.,* p. 4.
17. James E. Vance, Jr., "California and the Search for the Ideal," *Annals of the Association of American Geographers,* 62 (1972), 185-210.
18. *Ibid.,* 200.
19. David Lowenthal and Hugh C. Prince, "The English Landscape," *The Geographical Review,* 54, No. 3 (July, 1964), 309-346, and "English Landscape Tastes," *The Geographical Review,* 55, No. 2 (April, 1965), 188-222.
20. David Lowenthal and Hugh C. Prince, "English Façades," *Architectural Association Quarterly,* 1, No. 3 (July, 1969), 50-64.
21. David Lowenthal, "The American Scene," *The Geographical Review,* 58, No. 1 (January, 1968), 61-88.
22. *Ibid.,* 62.
23. See, for example, Peter Blake, *God's Own Junkyard,* Holt, Rinehart and Winston, Inc., New York, 1964, and Ian Nairn, *The American Landscape: A Critical View,* Random House, New York, 1965.
24. Pierce F. Lewis, David Lowenthal, and Yi-Fu Tuan, *Visual Blight in America,* Commission on College Geography Resource Paper No. 23, Association of American Geographers, Washington, D.C., 1973.
25. Pierce F. Lewis, "The Geographer as Landscape Critic," in Lewis, Lowenthal, and Tuan, *op. cit.,* p. 1.
26. Yi-Fu Tuan, "Visual Blight: Exercises in Interpretation," in Lewis, Lowenthal, and Tuan, *op. cit.,* pp. 23-27.
27. Kenneth Clark, *Landscape Into Art,* Beacon Press, Boston, 1951.
28. Tuan, "Visual Blight: Exercises in Interpretation," in Lewis, Lowenthal, and Tuan, *op. cit.,* p. 26.
29. D. W. Meinig, "Commentary: Visual Blight-Academic Neglect," in Lewis, Lowenthal, and Tuan, *op. cit.,* p. 45.
30. W. H. Parker, *The Soviet Union,* Aldine, Chicago, 1969, pp. 172-175. This is one work in the series *The World's Landscapes,* edited by J. M. Houston.
31. See, for example, the great differences between Marshall I. Goldman, "The Convergence of Environmental Disruption," *Science,* 170 (October 2, 1970), 37-42, and William M. Mandel, "The Soviet Ecology Movement," *Science and Society,* 36, No. 4 (Winter, 1972), 385-416.
32. Hazel Erskine, "The Polls: Pollution and Its Cost," *Public Opinion Quarterly,* 36, No. 1 (Spring, 1972) 120-135, quotation from p. 120. See also the second section of the discussion by the same author "The Polls: Pollution and Industry," *Public Opinion Quarterly,* 36, No. 2 (Summer, 1972), 263-280.
33. These figures are taken from W. R. Derrick Sewell and Harold D. Foster, "En-

vironmental Revival: Promise and Performance," *Environment and Behavior*, 3, No. 2 (June, 1971), 125. The entire issue is devoted to environmental quality.

34. This is discussed in James McEvoy III, "The American Concern with Environment," in William R. Burch, Jr., Neil H. Cheek, Jr., and Lee Taylor, eds., *Social Behavior, Natural Resources and the Environment*, Harper & Row, New York, 1972, pp. 214-236.
35. Denton E. Morrison, Kenneth E. Hornback, and W. Keith Warner, "The Environmental Movement: Some Preliminary Observations and Predictions," in Burch, Cheek, and Taylor, eds., *Social Behavior, Natural Resources, and the Environment, op. cit.*, pp. 259-279.
36. *Ibid.*, p. 272.
37. David Lowenthal, "The Environmental Crusade: Ideals and Realities," *Landscape Architecture*, 60, No. 4 (July, 1970), 290-346. See also by the same author, "Earth Day," *Area*, No. 4 (1970), 1-10.
38. See Footnote 33.
39. A thoughtful group of essays on national character is contained in Don Martindale, ed., "National Character in the Perspective of the Social Sciences," *The Annals of the American Academy of Political and Social Science*, 370 (March, 1967), entire issue.
40. Wilbur Zelinsky, *The Cultural Geography of the United States*, Prentice-Hall, Englewood Cliffs, N.J., 1973, p. 40.
41. *Ibid.*, p. 41.
42. *Ibid.*, p. 59.
43. *Ibid.*, p. 61.
44. Don Martindale, "Preface," in "National Character in the Perspective of the Social Sciences," *op. cit.*, x.

Eight

The world

"To traverse the world men must have maps of the world. Their persistent difficulty is to secure maps on which their own need, or someone else's need, has not sketched in the coast of Bohemia."
Walter Lippmann

Beyond the national level lie even broader perspectives of vast world regions or of the entire earth. Decision-making with consequences at the international level, as at all the other scales, is often based on subjective views rather than reality. But at the world scale our images of reality are bound to be faulty at best. Conceptions of other nations or world systems are based on the broadest of generalizations, to which many exceptions are inevitable. No one can understand more than a fraction of the earth's diversity at any one time. Therefore, our images are based largely on information provided by other people. There is great difficulty in assessing the degree of truth or error in such information, for it involves broad generalizations about persons, places, and events beyond the individual's own limited sphere of action. At no other scale is there less opportunity for reality-testing by the individual or group. Yet the consequences of collective ignorance are potentially more dangerous than at any other scale. It is important then to try to determine the kinds of conceptions of the world upon which such decisions are based.

This chapter will review some of the efforts to examine how people perceive the world or large portions of it beyond the scale of the individual nation. We will examine in turn national stereotypes, the mirror-image

phenomenon, the image in foreign policy decision-making, our geographical horizons, the Chinese perception of a world order, student views of the world, and Western attitudes toward nature.

Images of other nations

Every nation harbors images of other nations—in some cases rich and varied images based on people's experience, reading, and empathetic understanding. But more commonly the images are merely stereotypes, simplistic, inaccurate, and often unkind. The dangers of simplistic notions are recognized in the preamble to the constitution of the United Nations Educational, Scientific and Cultural Organization (UNESCO), which contains the observation: "Since wars begin in the minds of men, it is in the minds of men that the defenses of peace must be constructed."

The remarks of a Dutchman, Johan Huizinga, written the same year the Nazis invaded Holland, eloquently express his feelings about the tragic nature of international war.

> Every cultured and right-minded person has a particular affection for a few *other* nations alongside his own, nations whose land he knows and whose spirit he loves. Summon up an image of such a nation, and enjoy it. . . . It is not necessary to revisit the countryside or go to museums or to reread the poets. In your meditations you can lump all the treasures of that foreign nation together in one view. You perceive the beauty of its art, the vigorous forms of its life, you experience their perturbations of its history, you see the enchanting panoramas of its landscapes, taste the wisdom of its words, hear the sounds of its immortal music, you experience the clarity of its language, the depth of its thought, you smell the scent of its wines, you sympathize with its bravery, its vigor, its freshness, you feel all that together stamped with the ineradicable mark of that one specific nationality that is not yours. . . . All of this is alien to you—and tremendously precious as a wealth and a luxury in your life. Then why controversy, why envy?[1]

Unfortunately, the proportion of any population that can summon up such rich imagery of a foreign nation is far too small. Public surveys in the United States on topics of world affairs tend to reveal an abysmal ignorance of current problems and of the characteristics and locations of foreign nations.[2] The same is probably true in other nations.

An early attempt to investigate some of the ideas held by people of one nation concerning their own and other nations was a work by William Buchanan and Hadley Cantril, *How Nations See Each Other.*[3] As a major innovation, they tried to carry out surveys asking the same questions at approximately the same time of people in different nations. Under the auspices of the UNESCO study, "Tensions Affecting International Under-

standing," they carried out a cross-national survey, using modern sampling methods, in nine separate nations shortly after World War II.

To investigate national stereotypes, they used a technique previously developed by social psychologists in the United States to study stereotypes of races and nationalities. Each respondent was asked to select from a list of words those they thought best described their own people and people from a series of other nations.

The consistency with which various words were selected for each nation suggested that "stereotyped views of certain people are common property of the Western culture rather than the effect of bilateral national outlooks that differ from one country to another."[4] This may be seen from Table 8.1, which contrasts the Russian stereotypes, largely negative, with the descriptions of their own countrymen, invariably flattering.

Comparison of the scores on the survey with similar scores made in previous periods showed some marked shifts in the tone of stereotypes for various nations. Thus in 1942 Americans had a positive stereotype of the Russians that by 1948 had become rather negative. This seems to indicate that "stereotypes are less likely to govern the likes and dislikes between nations than to adapt themselves to the positive or negative relationship based on matters unrelated to the images of the people concerned."[5] The prevalence of complimentary over derogatory terms in a national stereotype might then be used as a good index of the friendliness between nations, particularly if a standard basis for comparison were established over a series of periods.

Harold Isaacs made an excellent study of the ideas and images Americans held of China and India.[6] Instead of word lists, however, he conducted long interviews designed to elicit the fullest possible recall of opinions, associations, memories, ideas, and experience. Those interviewed, 181 in total, were selected because they played an important or significant role in the communication process.

There were strong differences in the images of Chinese and Indians held by Americans. In general, there was a predominantly positive or admiring view of the Chinese, based on nearly two hundred years of contacts. A rich imagery had grown up, containing both positive and negative elements, which might be highlighted differently depending on changing circumstances. Thus the shift of emphasis following the Nixon trip to China could easily be made because of the backlog of positive views already available to Americans. In contrast, the American view of India is more predominantly negative. A major factor appears to be a negative reaction to the Indian caste system. Furthermore, Americans do not have the same richness of association for India that they have for China. Still, positive elements are present, such as a commonly expressed admiration for the quality of Indian religious thought.

The contrast between the generally positive views of China and the

Table 8.1 The Six Adjectives Most Frequently Used to Describe Five Nations (Brackets indicate tie in percentages.)

Description of Russians by

Australians	*British*	*French*	*Germans*
Domineering	Hardworking	Backward	Cruel
Hardworking	Domineering	Hardworking	Backward
Cruel	Cruel	Domineering	Hardworking
Backward	Backward	Brave	Domineering
Brave	Brave	Cruel	Brave
Progressive	Practical	Progressive	Practical
	Progressive		

Italians	*Dutch*	*Norwegians*	*Americans (U.S.)*
Backward	Cruel	Hardworking	Cruel
Cruel	Domineering	Domineering	Hardworking
Domineering	Backward	Backward	Domineering
Hardworking	Hardworking	Brave	Backward
Brave	Brave	Cruel	Conceited
Intelligent	Progressive	Practical	Brave
Progressive			

Description of own countrymen by

Australians	*British*	*French*	*Germans*
Peace-loving	Peace-loving	Intelligent	Hardworking
Generous	Brave	Peace-loving	Intelligent
Brave	Hardworking	Generous	Brave
Intelligent	Intelligent	Brave	Practical
Practical	Generous	Hardworking	Progressive
Hardworking	Practical	Progressive	Peace-loving

Italians	*Dutch*	*Norwegians*	*Americans (U.S.)*
Intelligent	Peace-loving	Peace-loving	Peace-loving
Hardworking	Hardworking	Hardworking	Generous
Brave	Intelligent	Brave	Intelligent
Generous	Progressive	Intelligent	Progressive
Peace-loving	Brave	Generous	Hardworking
Practical	Practical	Progressive	Brave
Conceited	Self-controlled		

SOURCE: William Buchanan and R. Cantril, *How Nations See Each Other,* University of Illinois Press, Urbana, 1953, pp. 51-52. Reprinted by permission.

essentially negative images of India may be associated with the degree of knowledge. An excellent cross-cultural study by Wallace Lambert and Otto Klineberg indicates that children appear to be better informed about those countries they find attractive than about those they find unattractive.[7]

The Lambert and Klineberg study, which provides many new insights into the development of stereotypes about foreign peoples, was the result of a survey of some 3,300 children of three different age levels from eleven parts of the world. The six-year-olds responded less frequently than the older children when questioned about foreign peoples. Their responses tended to be nonevaluative descriptions of facts, or general references to the good or bad qualities of the people in question. The older children

showed a larger repertoire of evaluative distinctions. The content of the descriptions shifted from physical features, clothing, language, and habits to personality traits, habits, politics, religion, and material possessions. In other words, there was a shift from comparison of observable and objective characteristics to more subtle, subjective features.

Children apparently come to think about foreign peoples in an increasingly more stereotyped manner between the ages of six and fourteen. Interestingly, the first signs of stereotyped thinking turn up in descriptions children give of their own group. The process of establishing the concept of one's own group tends to produce an exaggerated and caricatured view of one's own nation and people. Later learning may mark foreign groups as people who are different. Whether the people who are different will be regarded in a friendly or unfriendly fashion depends on the distinctive cultural differences used in training the children. The Lambert and Klineberg study indicated many variations in children's attitudes toward foreign peoples in the several nations surveyed.

The mirror-image phenomenon

Good and bad elements may be present in any nation's image of another nation. But in a situation of conflict the tendency is to overlook any favorable elements in the opponent's image. A decided lack of empathy develops. There seems to be a tendency to exaggerate the virtues of one's own side and the diabolical character of the opposite side. One's own group is seen as virtuous, and the opposing group, evil. However, the opposing group often has an exactly opposite image. What is black and white in one group's image becomes white and black to the other group. This has been labeled "the mirror-image phenomenon." It provides a striking example of the degree of distortion commonly found in people's perceptions at the international scale.

Ralph K. White has investigated the mirror-image phenomenon in the context of the conflict between the Soviet Union and the United States.[8] His findings, which seem to be in complete agreement with those of Urie Bronfenbrenner,[9] are based on interviews with former Soviet citizens outside the U.S.S.R. and with citizens inside the U.S.S.R. Basic to the Soviet citizen's image are the perception of the Soviet Union as peaceful and the perception of the United States as threatening.

That the Soviet Union is peaceful seems self-evident to Soviet citizens. It is easy for them to recall the horrors of World War II, when their country was devastated and their population decimated. To them it seems obvious that after such an experience the Soviet Union must be one of the most peace-loving nations in the world. Their nation's very strong support of weapons, rocketry, and the military establishment is interpreted as neces-

sary defensiveness. Any aggressive acts by the Soviet Union, if known or recalled, tend to be interpreted in the same light.

On the other hand, the United States is perceived as threatening. While there appears to be a strong underlying feeling of friendliness to America and Americans, most Soviet citizens seem to hold the view that the "rulers" of the United States are "threatening" them. The main things that disturb them are America's bases around their border, America's U-2 flights over the Soviet Union, and, above all, America's alliance with West Germany. Seen against the background of the assumed Soviet peacefulness, all these American actions are interpreted as clearly aggressive in intent.

In spite of the very great differences between the Soviet Union and the United States, there are surprising similarities in their images of themselves and each other. White has summarized some of these similarities:

> —The Soviet people tend to see their country as wholly peaceful. Americans tend to see theirs as wholly peaceful.
>
> —The vast majority of the ordinary Soviet citizens are not crusaders. They would not knowingly and willingly risk war for one moment in order to force their own ideology or way of life on others. Neither would Americans.
>
> —They are afraid. As they see it there is an enemy threatening them with a war that could become nuclear. Americans too see a threatening enemy.
>
> —They see not the common people but the rulers of the enemy nation threatening them with war. So do Americans.
>
> —Because they are afraid they endorse (at least to some extent) the power-seeking actions of their government. They arm, as they see it, in self-defense. So do Americans.
>
> —In their minds there is at least one great blind-spot; they cannot see that the West fears aggression by them. They cannot believe that Americans too (the American people as well as their leaders) are arming in self-defense. Similarly, many Americans cannot believe that the Soviet people really fear the West, or that defensive motives could exist along with aggressive ones in the minds of the Soviet leaders.[10]

The mirror-image concept implies nothing whatever as to the relative amount of truth on either side. The important thing is to recognize the psychological basis for such divergent images of the same situation. The Soviet and American examples reflect the general psychology of nationalism. Selective attention and slanted interpretations of events tend to perpetuate the images of one's own nation as heroic and virtuous. The white part of the black-and-white picture has a direct source in the strong, obvious, conscious desire of nearly all human beings to think well of their own group, which leads to interpretations of events consonant with that desire or wish. To maintain one's own image as a loyal citizen, a certain

amount of self-censoring takes place. There is also further pressure to conform to the beliefs of one's own group. Propaganda provides an additional impetus in aiding the mechanism of rationalization. But if we are not to blame, then they must be. Rationalization leads to the first part of the statement, "We are not to blame." The psychological mechanism of projection completes the picture. In a context of fear and suspicion, as when two nations are in conflict, it is easy to exaggerate to create an atmosphere of paranoid suspicion that fills in the black part of the picture. All that is required is to assume that the other side perceives "the world" as we do and acts on the basis of the same frame of reference. This is, of course, not true. But it is easier to grasp than the more complex truth that each group is acting in terms of its own conceptions of the situation.

Clearly, there are very strong psychological forces that help to perpetuate dangerous degrees of delusion at the international level. To overcome them, we need a more realistic estimate of our own nations's actions as well as a more empathetic view of the real world of others. Some might argue that the images held by ordinary people are of little relevance in international decision-making. However, there is little evidence that the political leaders are less subject to the general psychological characteristics of human beings than other people—a point recognized by political scientists who have attempted to incorporate a psychological dimension into their analyses of foreign policy decision-making.

The image in foreign policy decision making

The economist Kenneth E. Boulding was influential in pointing out to political scientists the importance of the image for foreign policy decision making. In his early book *The Image* he emphasized that it is what we think the world is like, not what is really is like, that determines our behavior.[11] In a pioneer paper linking elite images and foreign policy decisions he stated:

> . . . A decision involves the selection of the most preferred position in a contemplated field of choice. Both the field of choice and the ordering of this field by which the preferred position is identified lie in the image of the decision-maker.[12]

Thus to understand foreign policy it is essential to analyze the images of the decision makers, in terms of both their perceived field of choice and their evaluation of it.

Ole R. Holsti used the individual example of John Foster Dulles and his attitudes toward the Soviet Union to examine the role of images and their effect on foreign policy decisions.[13] His findings support the view expressed earlier that foreign policy decision makers share with the general public a tendency to interpret international incidents in light of their

personal belief systems. This idea will be explored further in connection with Israel's foreign policy.

An outstanding example of the role of images in foreign policy decision making is provided in Michael Brecher's book *The Foreign Policy System of Israel.*[14] Table 8.2 and Figure 8.1 represent the dynamic system through which inputs are channeled, processed, and transformed into outputs. In addition to many inputs from the real world, long recognized as important in foreign policy decision making, Brecher's scheme recognizes the importance of a psychological dimension. His framework provides a clear example of the distinction between the psychological and the operating environment at the scale of world political systems. In fact, this is the key element, according to Brecher, who states:

Table 8.2 The Research Design

Inputs		
Operational Environment		
External	Global	(G)
	Subordinate	(S)
	Subordinate other	(SO)
	Dominant bilateral	(DB)
	Bilateral	(B)
Internal	Military capability	(M)
	Economic capability	(E)
	Political structure	(PS)
	Interest groups	(IG)
	Competing élites	(CE)
Communication	The transmission of data about the operational environment by mass media, internal bureaucratic reports, face-to-face contact, etc.	
Psychological Environment		
Attitudinal Prism	Ideology, historical legacy, personality predispositions	
Élite Images	of the operational environment, including competing élites' advocacy and pressure potential	
Process		
Formulation	of Strategic and Tactical decisions in 4 issue areas:	
	Military-security	(M=S)
	Political-diplomatic	(P=D)
	Economic-developmental	(E=D)
	Cultural-status	(C=S)
Implementation	of decisions by various structures: Head of state, head of government, foreign office, etc.	
Outputs	The substance of acts or decisions	

SOURCE: Michael Brecher, *The Foreign Policy System of Israel*, Yale University Press, New Haven, Connecticut, 1972, p. 3. Reprinted by permission.

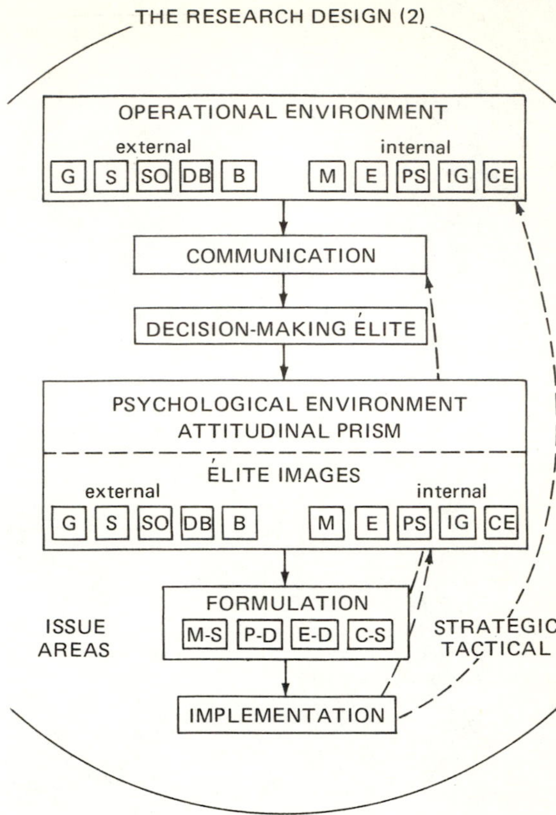

Fig. 8.1 Relationships of factors involved in foreign policy decision making. (Michael Brecher, *The Foreign Policy System of Israel,* Yale University Press, New Haven, Connecticut, 1972, p.4. Adapted by permission.)

The link between Image and Decisions is indeed the master key to a valuable framework of foreign policy analysis. This relationship of the two environments—operational and psychological—also provides a technique for measuring "success" in foreign policy decisions. To the extent that decision-makers perceive the operational environment accurately, their foreign policy acts may be said to be rooted in reality and are thus likely to be "successful."[15]

The operational environment includes both external and internal variables. *External variables* refer to all those conditions and relationships existing beyond the territorial boundaries of the state. These operate at several different levels, from the total global system to bilateral relations between any two states. Crucial to Israel at the intermediate level is the Middle East subordinate system seen in Figure 8.2. In addition to external

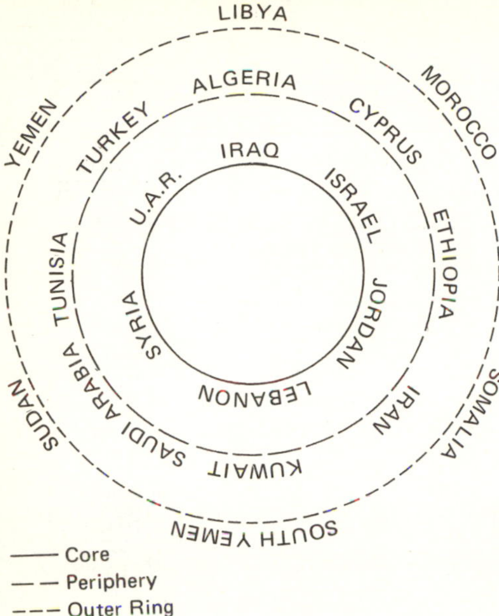

Fig. 8.2 The Middle East subordinate system from the perspective of Israel. (Michael Brecher, *The Foreign Policy System of Israel,* Yale University Press, New Haven, Connecticut, 1972, p. 49. Adapted by permission.)

variables are a series of *internal variables* that help define the character of the nation and its capabilities and limitations.

The inputs from the operational environment are filtered through the attitudinal prism, which is determined by both cultural factors and individual personality factors of the decision makers. This produces the elite images, the crucial inputs to the foreign policy systems. The reason for this is that the decision makers act in accordance with their perception of reality rather than in response to reality itself.

In light of his framework Brecher examines Israel's foreign policy within the global and regional systems of international politics during the first twenty years of Israeli independence. His analysis indicates several major achievements of Israel foreign policy as well as certain failures. The latter could be attributed to policy choices based on elite images. Since the failures stem more from the psychological than the operational environment, we will consider them further to illustrate the process.

Brecher's analysis led him to conclude that for the period as a whole there was a high degree of congruence between the image and the reality. A harsh, rigid, and forbidding reality was paralleled by a harsh, rigid, and forbidding perception of reality. The feedback from the psychological to the operational environment tends to create further rigidity and a poor

prognosis for resolving conflict. Change in either environment is difficult to generate, let alone sustain. This is particularly true if the attempt to induce change comes from only one side. Yet change must come to go beyond the present impasse. Brecher insists that the decision makers must perform what he considers a vital task of leadership, "self-analysis free of clichés."[16] His analysis reveals many examples of failure to perform this task adequately.

Brecher argues that Israel needlessly discarded a valuable power asset by too quickly and unequivocally aligning itself with the West in spite of early recognition by both the United States and the Soviet Union. This policy can only in part be explained by misjudgment of national interests. More important, according to Brecher, was the attitudinal prism of the "High Policy Elite," leaders, mainly of Russian origin, nurtured in the anticommunism of Social Democratic and Zionist ideology. This conscious choice of identification with the West may also have helped to increase Israel's isolation in the United Nations as the world body gained larger representation from African and Asian nations. The erosion of relations with the United Nations was also due to ambivalent, often derogatory perceptions of that body among Israel's decision makers.

Brecher also criticizes Israel's rejection early in 1955 of an approach from Peking to establish formal ties with China. He suggests that the direct cause was misperception, including misperceptions of the probable United States attitude and of the cost in economic and military aid. But it also reflected a general unfamiliarity with Asia in general and China in particular. Furthermore, the China decision indicated a misassessment of the importance of nonalignment. Israel's decision makers missed the opportunity to strengthen their natural link with the Third World. Their attitudinal prism led them to interpret the rebirth of Israel in terms of the historical and theological roots of the Jewish people instead of the global movement of independence. Failure to recognize or acknowledge the latter has led to isolation from the emerging Third World.

The supreme goal of peace in the Middle East has not yet been achieved, and the intransigence of the Arabs makes this goal exceedingly difficult to attain. But the Israeli leaders must also ask themselves what their behavior has contributed to this "diplomatic dead end." Brecher accuses the Israelis of "an inflexible adherence to the Ben Gurionist image of 'the Arabs' as Israel's implacable enemies, 'who understand only the language of force.' "[17] Although there is a large element of truth in this image, rigid adherence to it has led to a lack of innovation and imagination in foreign policy. "The qualitative jump in the Psychological Environment remains an historic task unfulfilled."[18]

Brecher's analysis of Israel's foreign policy system illustrates the important insights on international decision making that can be obtained when the image is emphasized. It would be valuable to have a whole series of such analyses made from the perspective of a wide range of individual

nations. In the next section we will examine some further images, not focused on the political dimension.

Our geographical horizons

An important aspect of international images is that of our geographical horizons—a topic speculated on some thirty years ago by Derwent Whittlesey in a stimulating essay that explored the way our geographic horizons have expanded over time.[19] He contrasted the primal, the regional, and the world sense of space.

The outlook of primitive people, as in the earliest period of human history, was almost literally confined to the physical horizon. The major landscape features might be known by their generic names, as "the mountain," "the river," and "the village." No doubt such narrowly restricted views of the world still survive, and even within modern cities there may be people whose minds only rarely touch on events beyond their block or neighborhood. According to Whittlesey, a broader regional sense of space appeared some 1,800 to 2,000 years ago as two areas of subcontinental size separately achieved political unity. The Romans and the Chinese, in producing a political unity based on effective internal communication, also provided some of their peoples with an awareness of a broader geographic horizon. Other regions appeared and former political regions contracted. But it was not until after the fifteenth century that a new, wider geographic horizon began to emerge. Following the European Age of Discovery, some of the areas beyond the oceans and the Sahara became known to Eurasian peoples. A worldwide horizon appeared for the first time at the most advanced frontiers of thought. Whittlesey felt it was reasonable to assume that eventually all human beings would advance to such a position, but his remarks that follow indicate clearly that he was sure this had not yet taken place at the time he wrote:

> Even in the lands where geography is part of a compulsory school curriculum, and among people who possess considerable information about the earth, the world horizon is accepted in theory and rejected in practice. A parochial outlook injects itself into every consideration of political system more inclusive than the one in vogue, it appears in snap judgements on "foreigners," and it dims the view of existing interdependence in the economic order. The myopia is understandable. The weight of tradition is heavy. Until five centuries ago a primal or regional sense of space dominated human settlements everywhere. In the United States, where society is mobile, it is easy to forget that even now few members of the human species move beyond their native region, and that hardly anybody has fruitful contact with folk who think in patterns different from his own.[20]

Clarence J. Glacken and John Kirtland Wright offer two examples of Western geographical horizons at different periods. Glacken's essay on the

Hellenistic Age (from the death of Alexander in 323 B.C. to the founding of the Roman Empire by Augustus in 30 B.C.), and its characteristic attitudes toward nature includes much information related to world views during a period of extraordinary culture contact.[21] Not only did the conquests of Alexander allow the Greeks to see entirely new lands and peoples, but even the old and familiar appeared in a new light. The world view was extended and changed, as in other great eras of change in Western history such as the Renaissance and the European Age of Discovery. In contrast, geographical horizons appeared to contract somewhat in the stagnant period preceding the time of the Crusades.

An excellent account of the geographical horizons of Western Europe just before the European Age of Discovery is provided in Wright's *The Geographical Lore at the Time of the Crusades.*[22] The book was first published in 1925, but one can see from the definition of its scope that Wright was a couple of generations ahead of his time in recognizing the importance of key concepts underlying the current interest in environmental perception and behavior.

> By "geographical lore" we mean what was known, believed, and felt about the origins, present condition, and distribution of the geographical elements of the earth. This covers a wider field than most definitions of geography. It comprises theories of the creation of the earth, of its size, shape, and movements, and of its relations to the heavenly bodies; of the zones of its atmosphere and the varied physiographic features of air, water, and land; finally it comprises theories of the regions of the earth's surface. Because many of these theories were false they are no less deserving of attention. The errors of an age are as characteristic as the accurate knowledge which it possesses—and often more so. Moreover, in addition to formulated beliefs, whether true or false, our definitions of geographical lore covers man's spiritual and esthetic attitude toward various geographical facts, as revealed—often unconsciously—in descriptions of regions or of landscapes.[23]

The intermingling of fancy and fact in the geographical lore at the time of the Crusades is clearly evident now. But, to the people of the period, it was no easy task to distinguish where fact faded off into fancy or fable. A major source of error was the excessive reliance on authority and tradition rather than the geography of observation. Wright conceived of the regional geographical lore of the age as consisting of two concentric circles, an inner circle that included those lands known firsthand through actual travel and an outer circle that included all the lands known secondhand through literary sources. The inner circle included essentially Europe west of the Elbe and Hungary together with the adjacent Mediterranean and Black Sea coasts. Beyond this small portion of the earth's surface was a vast, vague region of rumor and myth where fabulous monsters and legendary personages might be found. Even within the well-known zone were "islands of doubt and mystery," which, in the absence of information, could be

peopled by products of imagination. The outer semimythical zone gradually faded away as reliable observations replaced myths. But islands of ignorance still persist in most people's mental maps of the world. Here the distorted, vague, erroneous, and stereotypic notions of hearsay information may fill the vacuum resulting from lack of knowledge.

The persistence of imaginative misinformation in the absence of knowledge is well illustrated by a paper by Annemarie de Waal Malefijt, "Homo Monstrosus."[24] She points out that for 2,000 years most educated people believed that remote areas were inhabited by monstrous races such as those illustrated by Figure 8.3. In the eighteenth century Carl von Linné (Linnaeus) included in his classification of natural things the species Homo Monstrosus for the humanoid creatures with weird characteristics believed to exist in remote areas. It was not until the nineteenth century that it became clear that there was only one species of living human beings. But old myths die hard. Even today Homo Monstrosus survives in the occasional reports of an "abominable snowman" living, as might be expected,

Fig. 8.3 Monstrous tribes believed to inhabit remote lands. (From *Liber Chronicarum* by Hartmann Schedel, published in 1493.)

in one of the lesser-known areas on the earth, the remote mountain ranges of the Himalayas. The same lack of information and distortion of remote areas shows up in the following discussion of the Chinese perception of a world order and in some research on current images of the world. The application of Wright's scheme of a zonation in the quality of regional knowledge will be pursued further in the concluding chapter.

The Chinese perception of a world order

The Western world view is not shared by all the other inhabitants of the earth. Norton Ginsburg presents an example of an entirely different perspective.[25] His paper examines the Chinese perception of a world order. For centuries upon centuries the Chinese remained unchallenged as the central power in a vast Asian orbit. It is not surprising then that their traditional world view was strongly Sinocentric. Surrounding the core area of continuous Chinese control from ancient times was a series of roughly concentric zones in which their power and authority diminished gradually in all directions. Ginsburg distinguishes the four zones, illustrated in Figure 8.4. The Inner Asian Zone includes those areas not in the core area (Zone 1) but over which China has exerted various degrees of control, and with which China always had intimate relations. The Outer Asian Zone includes areas somewhat farther removed. These were relatively well known to the Chinese, though, with some exceptions, they were never in true tributary or client relationships with China. The Outer World (Zone 4) contained all the rest, areas that the Chinese were ignorant of or indifferent toward.

A vivid view of the type of information the Chinese had of one such area in Zone 4 is provided by Paul Wheatley's article on Chinese knowledge of East Africa before 1500 A.D.[26] At that time knowledge of foreign countries was very sketchy. Furthermore, such knowledge did not appeal to the Chinese public fancy. Wheatley states that the earliest known reference to East Africa in Chinese writings dates from 863 A.D. But not until much later did comprehensive views of broad areas appear, as in Figure 8.5. This represents East Africa as it would probably be envisaged by a well-informed Chinese official of the Sung (A.D. 1127-1279) or Yuan (A.D. 1280-1368) dynasty. There is no reason to believe that the accounts were based on personal experience, but by then a rudimentary knowledge was evident. This included a good understanding of the outline of portions of the coast, the predominantly Muslim port cities, and some notion of the Nile River and East African lakes. The source of knowledge, Arabo-Persian merchants and sailors, and the trading interests of the Chinese show up in such products noted as frankincense, the dragon's blood tree, and the whale. However, Chinese accounts of the time also included mythological ideas. Most notable was the fabulous p'eng bird, whose feathers were so

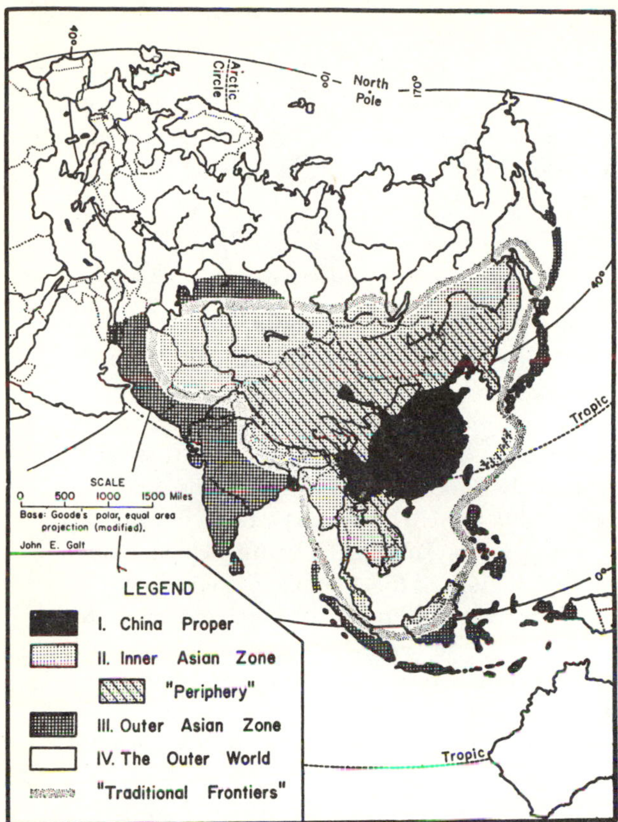

Fig. 8.4 China: The world order traditional model. (Norton Ginsberg, "On the Chinese Perception of a World Order," in *China's Policies in Asia and America's Alternatives,* ed. Tang Tsou, University of Chicago Press, Chicago, Vol. 2, 1968, p. 77. Reprinted by pemission.)

huge that a water jar could be fashioned from the quill. There was very little knowledge of the society along the African Coast, which is not surprising when we consider the attitude of conscious superiority cultivated by the Chinese.

The traditional Chinese view of the world has no doubt been altered by events of the past century. But it seems likely that many elements of the earlier view persist even today. Without awareness of such a world view it may be difficult to comprehend the major decisions in Chinese foreign policy. But if the world view is known, the foreign policy is much easier to understand—i.e., Taiwan is regarded as a part of the core area, and certain Soviet provinces were traditionally part of Zone 2.

The traditional Chinese view of the world is only one of many possible world views. The Israeli position within the Middle East system, seen in

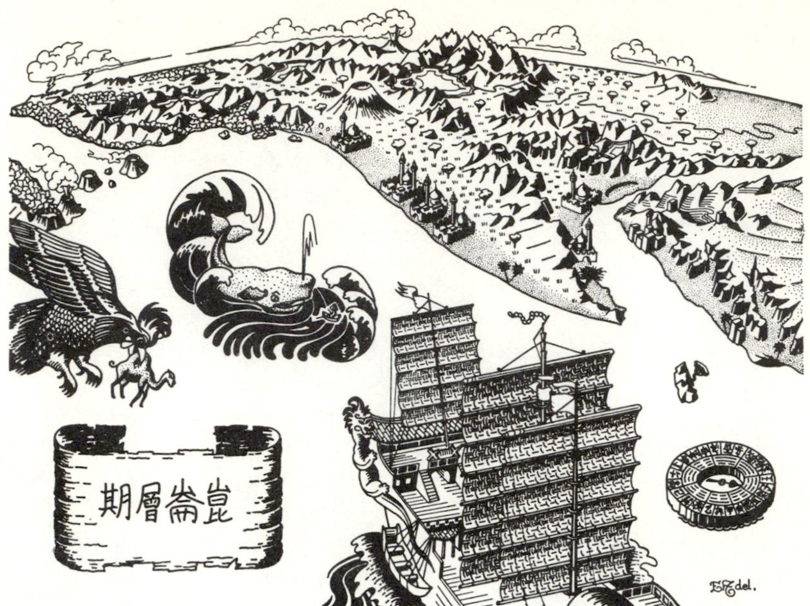

Fig. 8.5 East Africa as probably envisaged by a Chinese official of the Sung or Yuan Dynasty. (Paul Wheatly, "The Land of Zanji," in *Geographers and the Tropics: Liverpool Essays*, eds. R. W. Steel and R. M. Prothero, Longman Group Ltd., London, 1964, p. 158. Reprinted by permission.)

Figure 8.2, illustrates some of the dimensions of another world view. In each case there appears to be a central zone of greatest concern with a lesser degree of involvement in more distant areas. This same sort of pattern may be seen in the student views of the world that follow.

Student views of the world

The preceding discussion indicates that people have often had rather narrow geographic horizons dominated by a local or regional perspective. Does the same hold true today? To explore this question, I investigated student views of the world.

The method is simple and direct: Students were given a blank sheet of paper and asked to sketch a map of the world, labeling all the places they consider to be interesting or important. The maps obtained are fascinating, and they provide a direct means of investigating student views of the world.

The technique has been tried with several different groups of high school and university students from different parts of the world.[27] Students at this level generally have no problem producing a recognizable map of the world, although there are great individual differences in the amount of

information included. In spite of the differences there are always striking similarities seen in any set of sketch maps from the same place. The unit most commonly mentioned is the nation. This seems to be the most natural type of building block for a map of this scale. By calculating the frequency with which each nation is included, it is possible to construct a composite map for each group. Figure 8.6 shows an example, which represents the composite map for a group of high school students from Calgary, Alberta, Canada.

The composite map for Calgary illustrates a number of features common to most of the groups tested so far. The known and familiar areas tend to include the home continent and the world's largest nations, those of continental dimensions such as the United States, Brazil, the U.S.S.R., China, India, Canada, and Australia. For the Canadian students some of the Western European nations are also well known. The nations resemble Canada in being part of the wealthy, urban, industrial Western world. Nations more remote culturally—such as most of the nations in Africa, South America, and Asia—are not generally included. A notable exception among the Canadian students was the tendency to include unlike countries that share a link to the British Commonwealth of Nations.

A number of sketch maps by students from various parts of the world illustrate in extreme form some of the general tendencies of such maps (Figures 8.7 to 8.10). The tendency to exaggerate the importance of the home country and the surrounding region is obvious in all. This may be done by exaggerating its size, by placing it in a central position, or by including much more detailed coverage of features in the home area. In the maps included by students from the United States, Mexico, Sierra Leone, and Iran, the home country is clearly larger than life size. This is especially true when comparisons with other world areas are made. For example, Mexico is larger than Europe on the Mexican student's map, Iran is nearly as large as Africa on the Iranian's map, and Sierra Leone would be a giant country if transplanted full-sized to South America in the African student version of the world. The central position of North and South America common to many student sketch maps collected from Canada, the United States, and Mexico has unfortunate consequences for Eurasia, which is split in two to trail forlornly into the edges of the map, as in Figures 8.7 and 8.8. The greater detail in and around the home areas may be seen clearly in Figure 8.7, where the only United States state outlined is that from which the sample was drawn, and in Figure 8.9, where the countries bordering Iran appear in detail, although this entire region appears to be virtually unknown to the students from Canada, the United States, Mexico, Finland, and Sierra Leone sampled so far. The sole Iranian map was obtained on the first day of class from an Iranian student enrolled in introductory geography at the University of Arizona.

There is clearly much more diversity in world views than is apparent in the usual dichotomies such as developed-underdeveloped or communist-

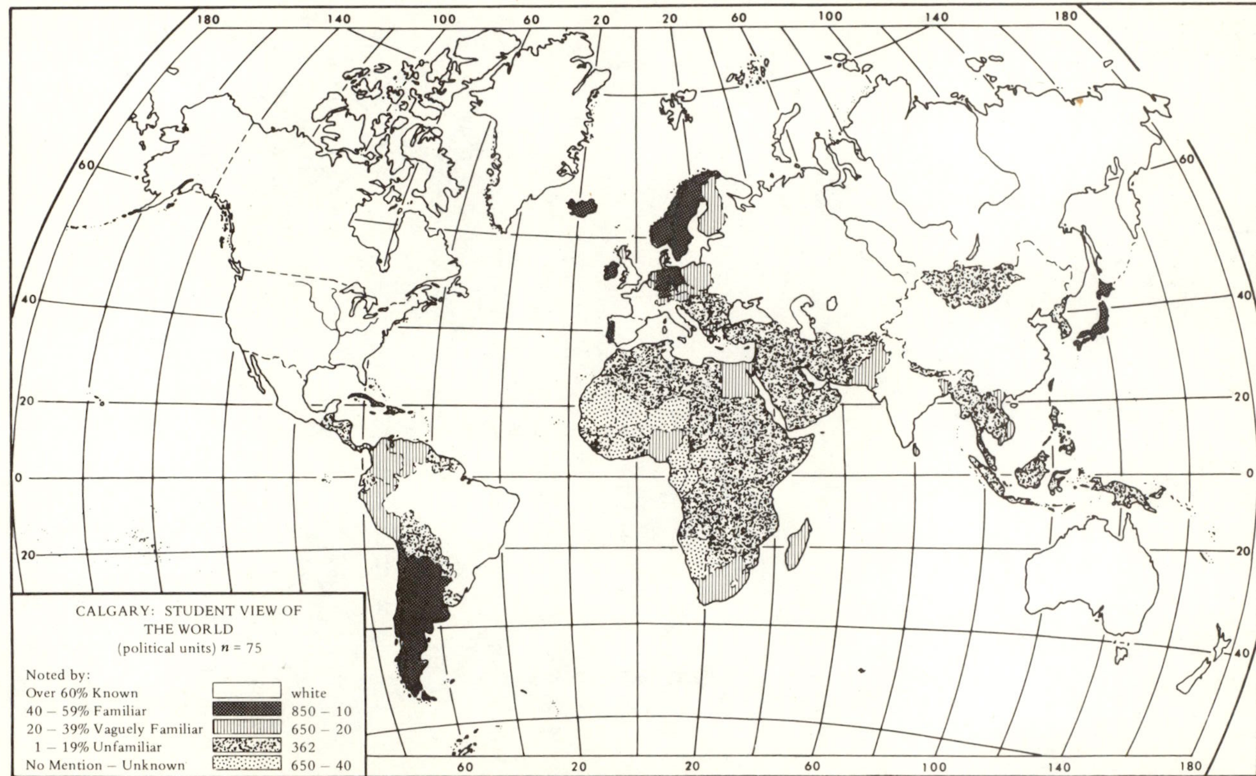

Fig. 8.6 Student views of the world composite map for students from Calgary, Alberta, Canada. (Thomas F. Saarinen, "Student Views of the World" in *Image and Environment*, eds. Roger M. Downs and David Stea, Aldine Publishing Co., 1973, p. 152. Adapted by permission.)

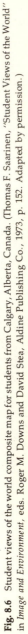

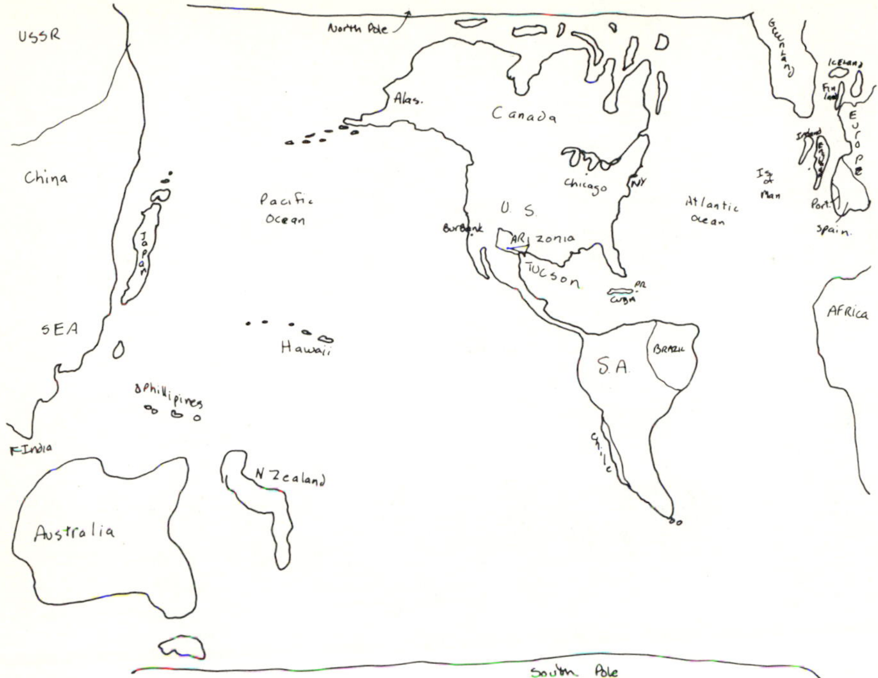

Fig. 8.7 An American student's sketch map of the world.

Fig. 8.8 A Mexican student's sketch map of the world.

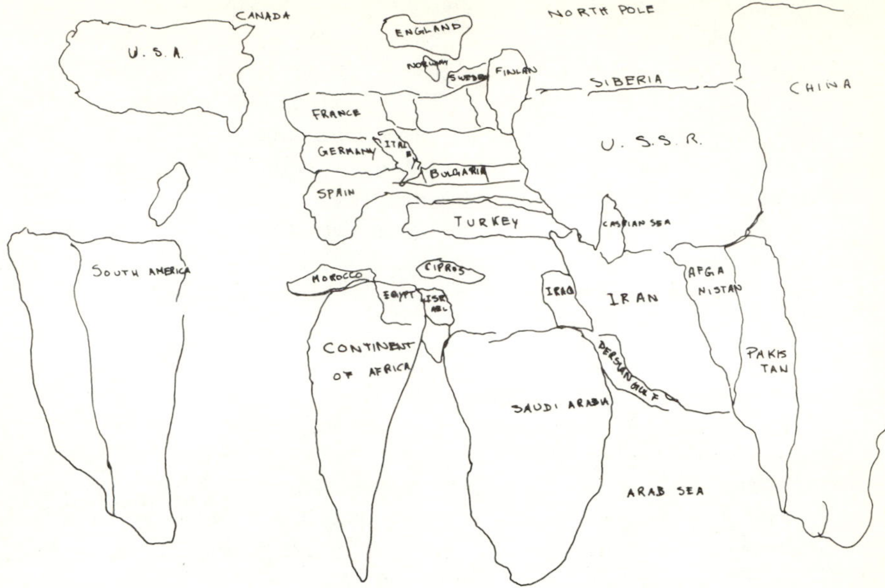

Fig. 8.9 An Iranian student's sketch map of the world.

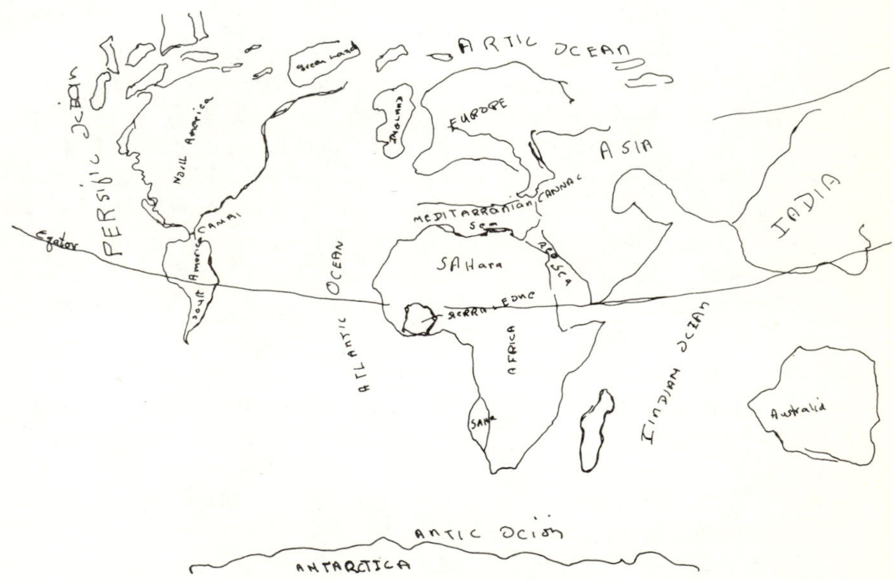

Fig. 8.10 A Sierra Leone student's sketch map of the world.

capitalist. Bruce M. Russett's analysis of United Nations voting patterns, for example, revealed six different coalition groups.[28] These he labeled the Western Community, Brazzaville Africans, Afro-Asians, Communist Bloc, Conservative Arabs, and Iberians. The similarity of the voting patterns of each group may reflect some broad commonly shared regional views of the world, such as the Chinese perception of a world order noted above. Kenneth E. Boulding has suggested that, to avert dangers in our unstable international system, more sophisticated world images are required.[29] Basic to the development of such images is an understanding of the variations in world views that exist currently and the way in which they are formed.

The school's role in forming world images

In discussing the development of stereotypes, we noted that they become fully developed only after several years of schooling. This fits the general pattern of political socialization investigated in a study carried out by Robert Hess and Judith V. Torney.[30] In an effort to determine how and when political attitudes are inculcated, they did research involving over 17,000 elementary-school children in the United States. The research was designed to gather information about early involvement in political affairs and to describe its nature and the general course it follows from early childhood into the adolescent years. The types of attitudes, concepts, and beliefs they studied are those involved in images of the world system of nations. A major conclusion was that "the school stands out as a central, salient, and dominant force in the political socialization of the young child."[31] Because the school's role is so dominant, they suggest that much more attention must be paid to evaluating the methods, curriculum, and timing of this political socialization.

Hess and Torney describe the strong positive attachment the child develops for the country as essentially an emotional tie that is exceedingly resistant to change or argument. There appear to be three stages in the development of the concept of the nation. In the first stage the focuses of attachment are national symbols, such as the flag and the Statue of Liberty. In the second stage the concept of nation is understood, including abstract qualities and ideological context. In the final phase the country is seen as part of a larger organized system of countries. The main concern of the Hess and Torney study was the political socialization into the system of the United States. But it seems likely that the same sort of process occurs in other countries—a process in which the children become imbued with a common cherished heritage, which has been referred to by the geographer Jean Gottman as their "iconography."

Gottman discusses iconography in an article examining the political partitioning of our world.[32] He states that "the real partitions, those which are most stable and least flexible, are in the minds of men. The worst

barriers stem from the diversity of the historical past."[33] Iconography is described as a powerful factor that resists change. It encompasses such things as the national flag, the proud memories of history, the principles of the prevailing religion, the generally accepted rules of economics, the established social hierarchy, the heroes quoted in the schools, and the classic authors. To be acceptable, any political changes that take place must be supported by special interpretations of the national iconography.

National education plays an important role in political socialization, and in so doing it may also pose a major obstacle to the creation of broad, empathetic world perspectives, because concepts of other nations may develop in contrast to our own. The more closely other nations resemble our own nation, the more likely they will be rated favorably, if our standards are assumed to be the best. Such biases show up clearly in elementary-school textbooks past and present.

An interesting analysis of nineteenth-century American schoolbooks is provided in Ruth Miller Elson's *Guardians of Tradition.*[34] In it she gives examples of the kinds of biases abounding in such literature. The differences between the American and European civilizations may be seen in the pictures heading sections of a schoolbook. America is represented by simple, honest citizens engaged in useful occupations, while Europe is represented by an aristocratic court, an avalanche, and the remains of a past civilization. Pictures of the races clearly communicate the superiority of the white race. One illustration, for example, has the white race represented by a refined and delicate woman, surrounded by rather swarthy, unpleasant men used to exemplify the other races. Another similar picture has the white race represented by the classically beautiful sculpture "Apollo Belvedere," while quite ordinary specimens are used for the other races.

Such blatant bias has been greatly remedied in modern textbooks, according to David Pratt,[35] who developed a method for measuring bias in textbooks. But he states further that studies conducted in the 1960s indicated that inaccuracies, omissions, and biases of a more subtle but no less pernicious kind could invariably be found. He provides examples from Canadian texts that inform the student that black immigrants are turning parts of British cities into "colored slums," that African nations tend to forget that there is still a great deal to be learned before they can take care of their own people, and that in many ways Asia must blame itself for its "backwardness." "Savage" is used to describe American Indians.

The perceived relative importance of various peoples and places may also be measured by the amount of space devoted to them. By this gauge the blacks in America do not seem to merit much attention.[36] Similarly, on a world scale European and American examples abound, while little attention is focused on the poorer nations of Asia, Africa, and South America. A major problem, at least in the United States, is that the biases of the textbooks are unlikely to be corrected by the teachers. For in their training

the teachers generally do not have courses on cultures other than their own,[37] let alone any meaningful contact with people from such cultures.

No universal history of human development has yet been written, and many conflicting accounts of events from differing nationalistic and geographic perspectives will have to be reconciled before one is possible. But it is apparent that the type of history taught today will have to be drastically altered to fit people for a world where all culture groups are now in contact. Newer, broader, more inclusive perspectives will be required. New symbols to help us extend our loyalties to the entire human race and to our home on planet earth must be created. We must seek consensus on people-environment attitudes and ideas that extend beyond the boundaries of single-nation states. Examples of such ideas held in common by broad groups of nations may be seen in the following discussion on Western attitudes toward nature.

Western attitudes toward nature

The most important factors to emerge in discussions of environmental attitudes beyond the national level tend to be ideological in nature. Those considering the people-environment relationship at this scale seem to see more clearly the practical consequences of religious or philosophical ideas.

A clear example is provided by Lynn White, Jr., who states that "Human ecology is deeply conditioned by beliefs about our nature and destiny—that is, by religion."[38] He contends that we can best understand the current ecological crisis in terms of its roots in the Judeo-Christian tradition. Christianity, in contrast to other religions, places us apart from nature and insists that it is God's will that we have dominion over nature. This mandate provides Christians with an uninhibited stance toward destructive exploitation of the environment, with the unfortunate consequences we now see developing around us. White argues that, since the roots of the ecological crisis are religious in nature, the remedy must also be essentially religious. He recommends adoption of more loving and reverent attitudes toward nature, such as those exemplified by Francis of Assisi, who could well serve as the patron saint of ecologists.

Although presenting a more complex analysis of Christian attitudes toward nature, John Black, as White, indicates that an understanding of such ideas is a prerequisite for an intelligent analysis of our present situation.[39] He presents in detail a countervailing view, which also has deep roots in the Judeo-Christian tradition. This is the concept of "stewardship," which reflects the ideas of responsibility and trust. In this view God is pictured as an absentee landlord of the earth and its resources, with the human being as His "steward." Such imagery evokes a long-range responsible stance. However, in Western civilization the belief in the right to control nature remains dominant. In fact, it has intensified with the

availability of more advanced technology and increasing pressure of population on resources.

Black argues, however, that the present ecological situation should not be considered a crisis but a state of continuous transition. To make any progress in rational use of the world's resources, a long-term perspective of at least 250 or 500 years should be adopted. This requires a new vision of ourselves, and a new world view, which, to be successful, must be based on widely accepted ideas within the contemporary world view. Black suggests that this might best be based on what he considers the most firmly fixed of all Western values, the linear concept of time.

> By expanding our view of the general good of "mankind" in terms of the whole of humanity, dead, living or as yet unborn, we may perhaps be able to assess what we do in terms of the good of mankind, regardless of the position of the individual along the time axis of the world.[40]

A recent attempt to develop new principles of behavior and responsibility may be seen in the Declaration on the Human Environment emanating from the Stockholm Conference.

The Stockholm Conference of June, 1972, dramatically emphasized the worldwide growth of environmental awareness and the increasing recognition that environmental problems transcend national boundaries.[41] For the first time the representatives of 110 governments convened in an effort to cope with environmental problems on a comprehensive and global basis. As the preparations for the conference proceeded, it became apparent that there were major differences in outlook between the rich and poor nations. In Chapter 7 we noted that within the United States a greater concern for environmental quality has been associated with the most affluent segments of society. Similarly, on a global scale representatives of the rich nations expressed concern about deterioration in environmental quality. Those from the poorer nations had a different perspective, one that indicated that the environmental problem requires a new view of our relationships not only with the natural world but with other people. As Maurice Strong expressed it:

> We have the representatives of the poor of the world to thank, then, for expanding both Stockholm's and the world's environmental awareness to enhance not just affluence and wastes, the dangers, disamenities, and aesthetic insults found in industrialized societies, but the malnutrition, disease, illiteracy and human degradation which are the dominant characteristics of the human environment for most of mankind.[42]

Although most authors who consider the relationship between people and environment at the scale of the world emphasize the importance of ideology, it should not be assumed that there is a simple, one-to-one relationship between environmental attitudes and behavior. As Yi-Fu

Tuan has correctly pointed out, there are often discrepancies between the stated ideals of a group and the reality.[43] Reacting to White's thesis noted previously, he observes that lack of the Judeo-Christian ethic of dominion over nature did not prevent the Greeks and Romans from drastically modifying their environments. Nor did the Chinese landscape escape degradation, in spite of that culture's ancient tradition of an adaptive attitude toward nature. Clearly, any culture offers many possible philosophies to choose from, and particular circumstances may influence which one gains ascendancy.

An appreciation of the deep roots and rich variation in ideas that underlie Western thought on people-environment relations may be gained from Clarence Glacken's *Traces on the Rhodian Shore.*[44] Here he traces three general ideas from classical antiquity to the eighteenth century—the idea of an earth created by design, the influence of environment on human beings, and the influence of human beings on their environment. The early history of these ideas helps make more meaningful modern attitudes toward the earth. For example, we can see that much of the discussion of ecology today is deeply rooted in the idea of a designed earth, which involves an effort to understand environments as wholes, as manifestations of order. Similarly, the problems considered under the general label of "ecological crisis" are manifestations of our influence on our environment, which has been observed and discussed throughout the ages. The idea of the influence of the environment on people remains a lively issue in all discussion concerned with designing better buildings, neighborhoods, and cities. Glacken's work invites comparison of Western thought on this subject with that of other major traditions, like the Indian, Chinese, and Moslem.

The foregoing studies on Western attitudes toward nature have illustrated the concern with ideologies at that level of generalization. As one goes beyond to compare Western thought to that of other cultures, symbols become supreme. This may be seen in Yi-Fu Tuan's monograph *Man and Nature*[45] and his later book *Topophilia,*[46] in which he discusses symbolic meanings of such landscapes as the wilderness, the garden, and the city. His search for universal human themes leads to an analysis of symbolism associated with our biological nature and the types of cosmologies developed in response to generalized types of physical environments.

Summary

At the world scale the transition from perceptions of the environment to conceptions or images of the environment is complete. No longer is the predominant concern focused on the real world. Instead, it is images and ideologies that dominate. Although some of the characteristics commonly

considered with smaller segments of the environment remain, there is a general shift to the symbolic aspects that lie behind the various images.

In the chapter on the world scale some by now familiar themes are once more touched on. The concern with cognitive maps is reflected in the student views of the world and in the many regional perspectives considered. The importance of the decision-maker is once more emphasized, as is the importance of examining the images that influence decisions. The need to consider the interactions among nations as a total system with interlocking subsystems bears a close resemblance to the systems approach advocated by other authors at other scales. Brecher's framework for analysis of Israel's foreign policy system is remarkably similar to the Downs scheme proposed as an appropriate general framework for the full gamut of studies on environmental perception behavior and planning. The overriding importance of culture appears again in the discussion of national iconographies and political socialization.

The role of the educational system is discussed in greater detail here than in previous chapters. The schools are, of course, important at all scales in heightening awareness of environmental problems. But the greatest need for a change in direction appears at the world scale. National education serves to unite the many different groups within a country to create loyalties to this larger unit. At the world level, however, it may lead to confrontations of incompatible ideologies. These have great power to stir the emotions of millions of people in ways that impede rational action and thus lessen survival possibilities.

At the global level environmental images are clearly more fantastic and less related to reality than at any other scale. Because of the lack of direct feedback from the environment, greater reliance is placed on stereotypes, such as those that appear in school textbooks or in the conceptions of other nations held by children. The remoteness of portions of the environment enabled people in the past to entertain such notions as "Homo Monstrosus." The same lack of direct contact also provides a suitable atmosphere for development of the mirror-image phenomenon. The enormous sums of money devoted to war industries indicate how real the fears arising from such images are. Educational systems that perpetuate such pathological fears are anachronistic in an age where the weapons of destruction are capable of destroying the earth. Yet, as Brecher's study indicates, nothing less than a deliberate change in the psychological environment is needed to arrest the momentum of past images.

Planning for the international system is at the opposite pole from the type of planning noted in the early examples of people-machine systems. The most crucial need is to work with the mental, especially the symbolic, dimensions of human beings. This is a much more difficult task, but one upon which our very survival may depend. Yet at this level, as at all the others, the same sort of parochial views prevail. These will be considered in greater detail in the final chapter, in which we will call for a new kind of

regional consciousness, one that unites all people in a concern for our home on earth.

Symbols and ideologies hold sway at the world scale. Whether one talks of the Stockholm Conference or views of other nations it is clear that behind each image is an ideology absorbed in one's culture. Thus much of the discussion of the world ecological problem tends to center on differences of ideology, which have practical consequences in terms of actions likely to be taken. While some of the ideologies transcend the nation, and thus may have some potential for bringing about unity, a world-wide consensus is still lacking. Black suggests that the failure to develop such an ideology may be considered the major failing of Western civilization.[47]

Notes

1. Johan Huizinga, *Man and Ideas*, trans. James S. Holmes and Hans van Marle, Meridian Books, Inc., New York, 1959, p. 154. I would like to thank my Hungarian colleague Dr. András Szesztay for bringing this quotation to my attention.
2. John P. Robinson, *Public Information About World Affairs*, Survey Research Center, Institute for Social Research, University of Michigian, Ann Arbor, Mich., 1967.
3. William Buchanan and Hadley Cantril, *How Nations See Each Other*, University of Illinois Press, Urbana, Ill., 1953.
4. *Ibid.*, p. 50.
5. *Ibid.*, p. 56.
6. Harold R. Isaacs, *Images of Asia: American Views of China and India*, Capricorn Books, New York, 1962 (published originally as *Scratches on Our Minds*, 1958).
7. Wallace E. Lambert and Otto Klineberg, *Children's Views of Foreign Peoples*, Appleton-Century-Crofts, New York, 1967, p. 215.
8. Ralph K. White, "Images in the Context of International Conflict: Soviet Perceptions of the U.S. and the U.S.S.R.," Chapter 7 in Herbert Kelman, ed., *International Behavior*, Holt, Rinehart and Winston, Inc., New York, 1965, pp. 238-276.
9. Urie Bronfenbrenner, "The Mirror Image in Soviet-American Relations," *Journal of Social Issues*, 17 (1961), 45-46, as noted in Ross Stagner, *Psychological Aspects of International Conflict*, Wadsworth Publishing Co., Belmont, Calif., 1967.
10. White, *op. cit.*, p. 255.
11. Kenneth E. Boulding, *The Image*, The University of Michigan Press, Ann Arbor, Mich., 1956.
12. Kenneth E. Boulding, "National Images and International Systems," *The Journal of Conflict Resolution*, 3, No. 2 (June, 1959), 120-131; quote at p. 121.
13. Ole R. Holsti, "Cognitive Dynamics and Images of the Enemy," *Journal of International Affairs*, 21, No. 1 (1967), 16-39. The entire issue is devoted to "Image and Reality in World Politics."
14. Michael Brecher, *The Foreign Policy System of Israel: Setting, Images, Process*, Yale University Press, New Haven, Conn., 1972.
15. *Ibid.*, p. 4.
16. *Ibid.*, p. 564.
17. *Ibid.*, p. 564.

18. *Ibid.*, p. 564.
19. Derwent Whittlesey, "Horizon of Geography," *Annals of the Association of American Geographers*, 35, No. 1 (March, 1945), 1-36.
20. *Ibid.*, 14.
21. Clarence J. Glacken, *Traces on the Rhodian Shore: Nature and Culture in Western Thought From Ancient Times to the End of the Eighteenth Century*, University of California Press, Berkeley, Calif., 1967, pp. 18-33.
22. John Kirtland Wright, *The Geographical Lore of the Time of the Crusades*, Dover Publications, New York, 1965.
23 *Ibid.*, p.1.
24. Annemarie de Waal Malefijt, "Homo Monstrosus," *Scientific American*, 219 (October, 1968), 112-118.
25. Norton S. Ginsberg, "On the Chinese Perception of a World Order," in *China's Policies in Asia and America's Alternatives*, Tang Tsou, ed., University of Chicago Press, Chicago, 1968, pp. 73-91.
26. Paul Wheatley, "The Land of Zanj: Exegetical Notes on Chinese Knowledge of East Africa Prior to A.D. 1500," in *Geographers and the Tropics: Liverpool Essays*, R. W. Steel and R. M. Prothero, eds., Longman, London, 1964, pp. 139-188.
27 Thomas F. Saarinen, "Student Views of the World," in *Image and Environment*, Roger M. Downs and David Stea, eds., Adline Publishing Co., Chicago, 1973, pp. 148-161, and "The Effect of the Border on Student Views of the World: Nogales, Arizona and Nogales, Sonora," paper presented at the Rocky Mountain Social Science Association Meeting, El Paso, Tex., April, 1974.
28. Bruce M. Russett, *International Regions and the International System: A Study in Political Ecology*, Rand McNally and Co., Chicago, 1967.
29. Kenneth E. Boulding, "National Images and International Systems," *Journal of Conflict Resolution*, 3 (1959), 120-131.
30. Robert Hess and Judith V. Torney, *The Development of Political Attitudes in Children*, Aldine Publishing Co., Chicago, 1967.
31. *Ibid.*, p. 219.
32. Jean Gottman, "Political Partitioning of Our World," *World Politics*, 4, No. 4 (July, 1952), 512-519.
33. *Ibid.*, p. 519.
34. Ruth Miller Elson, *Guardians of Tradition: American Schoolbooks of the Nineteenth Century*, University of Nebraska Press, Lincoln, Nebr., 1964.
35. David Pratt, *How to Find and Measure Bias in Textbooks*, Educational Technology Publications, Englewood Cliffs, N.J., 1972.
36. Fred Donaldson, "Geography and the Black American: The White Papers and the Invisible Man," *The Journal of Geography*, 70, No. 3 (March, 1971), 138-149.
37. Harold Taylor, *The World As Teacher*, Doubleday and Company, Inc. , Garden City, N.Y., 1970, pp. 21-24.
38. Lynn White, Jr., "The Historical Roots of Our Ecological Crisis," *Science*, 10 (March, 1967), 1203-1207.
39. John Black, *The Dominion of Man: The Search For Ecological Responsibility*, Edinburgh University Press, Edinburgh, 1970.
40. *Ibid.*, 123.
41. Department of State, *Documents For the U.N. Conference on the Human Environment: Stockholm, June 5-16, 1972, Part 2*, National Technical Information Service, Springfield, Va., 1972.
42. Maurice F. Strong, "One Year After Stockholm: An Ecological Approach to Management," *Foreign Affairs*, 5 (1973), 690-707; quote on p. 692.
43. Yi-Fu- Tuan, "Discrepancies Between Environmental Attitude and Behavior:

Examples From Europe and China," *The Canadian Geographer,* 12, No. 3 (1968), 176-191.

44. See Footnote 21.
45. Yi-Fu Tuan, *Man and Nature,* Commission on College Geography Resource Paper No. 10, Association of American Geographers, Washington, D.C.: 1971.
46. Yi-Fu Tuan, *Topophilia: A Study of Environmental Perception, Attitudes, and Values,* Prentice-Hall, Inc., Englewood Cliffs, N.J., 1974.
47. Black, *op. cit.,* p. 145.

Nine

Conclusions

"Reality is like anything we build; first we shape it, and then it shapes us."
Paul Tibbetts and Aristide Esser

At the broadest level the studies reviewed in the preceding chapters share the objective of aiding us in making a more harmonious adjustment to our environment. Environmental planning is based not so much on the environment as it is but, rather, on the environment as it is perceived. To understand how decisions are made and, ultimately, to improve the quality of environmental design, it is necessary to understand environmental perception and behavior. This includes identifying the people most involved, determining their behavior, needs, and preferences, and weighing these against the environmental possibilities.

Chapters 1 through 8 were organized by scale to illustrate the types of research at each level, as well as the changes in emphasis that occur at different-sized segments of the environment. We noted, for example, the changes in perceptions as the scale moved from smaller to larger units of the environment. Naturalistic observation was seen to be useful to study environmental behavior in smaller-sized areas, such as rooms and buildings, while a broad-sampling technique, such as public opinion polling, was used for large segments of the environment, such as nations.

This chapter will discuss some common themes that serve as links through the range of studies of personal space or room geography to the

world. A number of recurring observations indicate current weaknesses in environmental planning and suggest directions for future emphasis. Because so little is currently known about people's needs, preferences, and behavior in relation to the environment, design should be regarded as an experiment. Postdesign analysis of newly planned environments would allow adjustments to be made, and in this way environments ever more supportive of human needs and preferences could be created. A major reason for past failures in environmental planning is our strong tendency toward ethnocentric thinking. The sections in this chapter that discuss parochial viewpoints and zones of knowledge, professional perspectives on environmental perception, and the visual-semantic communication gap describe this problem. Clearly, we must cultivate more comprehensive perspectives.

Interlocking levels

If our world (spaceship earth) is to become a harmoniously functioning system, we must understand all the parts contributing to the whole. The arrangement of chapters in this book, ranging from personal space and room geography to the world, is based on the conviction that all are interrelated as parts of the same system. The smaller-sized units are subsystems of the larger units, although each system may function according to its own laws. Such an interlocking set of systems has been well described by Roger G. Barker:

> A frequent arrangement of ecological units is nesting assemblies. Examples are everywhere: in a chick and embryo, for example, with the cells of the organs, the nucleus of one of the cells, the molecular aggregates of the nucleus, the molecules of an aggregate, the atoms of one of the molecules, and the subatomic particles of an atom. A unit in the middle ranges of a nesting structure such as this is simultaneously both circumjacent and interjacent, both whole and part, both entity and environment. An organ— the liver, for example—is whole in relation to its own component pattern of cells, and is a part in relation to the circumjacent organism that it, with other organs, composes: it forms the environment of its cells, and is, itself, environed by the organism.[1]

Many studies we have cited illustrate the interlocking of different-sized units of the environment. Different furniture arrangements, for example, led to different behavioral patterns within a room. Zeisel's study showed the importance of relating the behavior in kitchens and living rooms to the total apartment layout. Newman's work on defensible space suggested that the arrangement of apartment buildings may determine the quality of neighborhood relations in urban areas. The ecological psychologists demonstrated the interlocking set of behavioral settings that make up a small

town. In his study of the city of San Cristobal, Wood described the replication of successful design at several different scales. He argued that this aided in the transfer of skills from lower to higher levels in the city's hierarchy of places. Although a city may be regarded as an independent system, it also has external relations that tie it to the surrounding region, to the nations, and to the rest of the world. Similar examples were provided at the world level, where Brecher's work illustrated how Israel's position within the Middle East Subordinate System interacted with global foreign policy decisions.

The main point is that to understand fully any unit of the environment, it is necessary to consider not only the units it contains but the larger unit in which it is contained. Thus we could think of the world as a system containing subsystems of nations. Each nation in turn consists of a subsystem of regions. The region may encompass only one city or perhaps a set of cities, and each in turn consists of a subsystem of neighborhoods, and so on to buildings, homes, and rooms within them.

An ecological approach

Each level within the hierarchy of systems is best studied by an ecological approach. In other words, what is required is a consideration of all the factors and of their relationships to one another as parts of a single system. This was first illustrated by the example of the astronaut in a space capsule—a sensitively attuned people-machine system. We discussed the concept of behavioral settings developed by the ecological psychologists. As in the case of the space capsule, we described a system that consisted of human components, nonhuman components, and control circuits. In this people-environment system we considered not only the physical dimensions but also the psychological dimensions. These too must be included in environmental design at all scales.

If a particular segment of the environment is operating as a system, one would expect that a change introduced at any point would have reverberations through the system. This type of effect was noted in many of the studies we cited.

The ecological crisis provides at the global and national scales innumerable examples of such a process. Air pollution, water pollution, and, in fact, all the environmental problems can be described as inadvertent side effects of our increasingly powerful interventions in the natural environment. Although we are far from having sufficient knowledge to trace the repercussions of our environmental interventions, the need to do so has been recognized, for example, in the United States legislation that requires environmental impact statements for major projects. The geographic work on perception of natural hazards is also being pursued at the

regional, national, and global scales. The ecological approach of these geographers aims to provide public policy decision-makers with the kind of information that should enable them to make more sensitive adjustments to the environment. The study of the extreme events caused by natural hazards serves in this case to highlight normal behavior in relation to more usual conditions.

Some intriguing examples of system effects were also seen in the smaller segments of the built and social environment. The subtle interactions between distance, eye contact, and conversation were noted in Chapter 2 on personal space. Sommer showed how changes in furniture arrangement not only increased the number of conversations between elderly women but also had the unanticipated result of increasing the amount of reading and craft activities. Similarly, the work of Ittelson, Proshansky, and Rivlin illustrated that the behavioral consequences of physical changes in one room were not confined to that room alone but extended throughout an entire psychiatric ward. While these systems effects have been observed, we do not understand fully how they come about. For this reason it has been suggested that, at the present stage of our knowledge, the most fruitful approach would be to regard any plan or design as an experiment.

Design as experiment

Ideally, design goals should be based on human needs, but this is not easy to do because needs are difficult to observe or measure. Raymond Studer argues that the critical need in environmental design is a unit of analysis with dimensions that are both relevant to the participants and readily observable.[2] He suggests that human behavior is the best unit of analysis because our needs can be only *inferred* from behavior rather than observed. By analyzing human behavioral systems, relevant problem spaces may be detected, isolated, and structured. These may differ from preconceived ideas of buildings, rooms, houses, or cities. The physical product must then be tested to determine its congruence with the goals and capacities of the participants.

> Designed environments, then, should be viewed as *experiments* in which relevant variables—either behavioral or environmental—are manipulated (either by the participants or others) to move the system toward a state of consonance with respect to the goal structure in effect.[3]

Design for human well-being should be considerd an iterative process, constantly modified and moving toward more appropriate conditions.

Studer's suggestions resemble Perin's that were considered in the summary of Chapter 3 in that both are concerned with providing environments that are congruent with the needs or goals of the participants. This requires

a fresh look at the total people-environment system at each scale and new conceptual tools that will enable us to grapple with the real world in all its complexity.

Planning that fails to consider the activities of the main participants in a particular segment of the environment has been seen to create more problems than it solves. Examples abound in the studies cited above—bathrooms designed for the convenience of the plumbers, furniture arranged for the convenience of the janitors, radiology departments designed to suit the equipment, and urban renewal projects that ignore the needs of the slum dwellers. Such partial planning is not at all unusual and could be considered a natural outcome of the parochial perspectives so characteristic of most of the public and professional people involved in environmental decision-making.

Parochial viewpoints and zones of knowledge

Strikingly parochial viewpoints commonly appear when people are asked to draw maps of regions or express preferences for places. Generally, the areas closest to the location of the individual are sketched more accurately in terms of shape, and greater detail is given them. Furthermore, the home areas tend to be preferred over all others. The central placement and exaggeration in size of features closest to home further illustrate their perceived importance. This seems to hold true regardless of the scale of inquiry. Studies in previous chapters showed this effect at the scale of the campus, the neighborhood, the city, the nation, and the world. Denis Wood provides an example from a block in the city of San Cristobal in the state of Chiapas, Mexico.[4] The sketch of the block façade in Figure 9.1 shows the individual's home as central and more detailed than the other houses in the block. Sommer notes a similar effect on a regional scale in recall of town names in Saskatchewan.[5] It seems likely that the gradient of information observed in these studies reflects roughly the amount of knowledge the individual has about regions of different scales, and that the amount of information retained is probably related to the degree of relevance to the individual. One probably learns as much as is necessary to function comfortably at each level, but little more.

The tendency toward parochial views has some similarities to the Wright scheme of concentric zones of regional geographic knowledge discussed in Chapter 8. Although Wright's scheme was originally conceived at a broad international level, it may be applied profitably at many other scales. An illustration of such an application may be seen in John Allen's analysis of the exploratory process.[6]

Allen selected the Lewis and Clark expedition of 1804-1806 to study the

Fig. 9.1 Student drawing of his block façade. (Denis Wood, *Fleeting Glimpses*, Clark University Cartographic Laboratory, Worcester, Massachusetts, 1971, p. 199. Reprinted by permission.)

relationship between the exploratory process and our understanding of our world. Gradations in the accuracy of regional knowledge were the bases of the assessment. The accuracy of regional knowledge was assessed in terms of the currently accepted geographical reality and in terms of how the information available was perceived by those who used it. Both the real knowledge and the perceived knowledge of a region can be zoned in terms of the quality of information available. This is illustrated in Figure 9.2. Allen explains:

> We can divide a region into zones of first-degree knowledge, obtained through active commercial, diplomatic, ecclesiastical, military, and scholarly enterprise; second-degree knowledge, derived from travelers' accounts and/or fairly reliable hearsay; and third-degree knowledge, acquired only through rumor and conjecture.[7]

Unfortunately, the perceived quality of the knowledge may differ substantially from the actual reliability of the knowledge, as Figure 9.2 shows. The first and second zones of perceived knowledge tend to extend beyond the first and second zones of real knowledge. In other words, the explorers think they know more than they actually do. Furthermore, the field operations are based on the perceived knowledge. It is only as the exploration progresses that the explorers come to realize discrepancies between the zones of actual and perceived knowledge. Gradually, as the inadequacy of the information becomes apparent, the explorers may shift from a reliance on geographical lore obtained before the exploration to detailed information obtained from the natives and from field observations during the course of exploration.

The success of the expedition may depend on the ability to recognize changes in the operational zones. Allen illustrates the process by referring to the Lewis and Clark expedition. He shows how new information changed the zones of both real and perceived knowledge and enabled

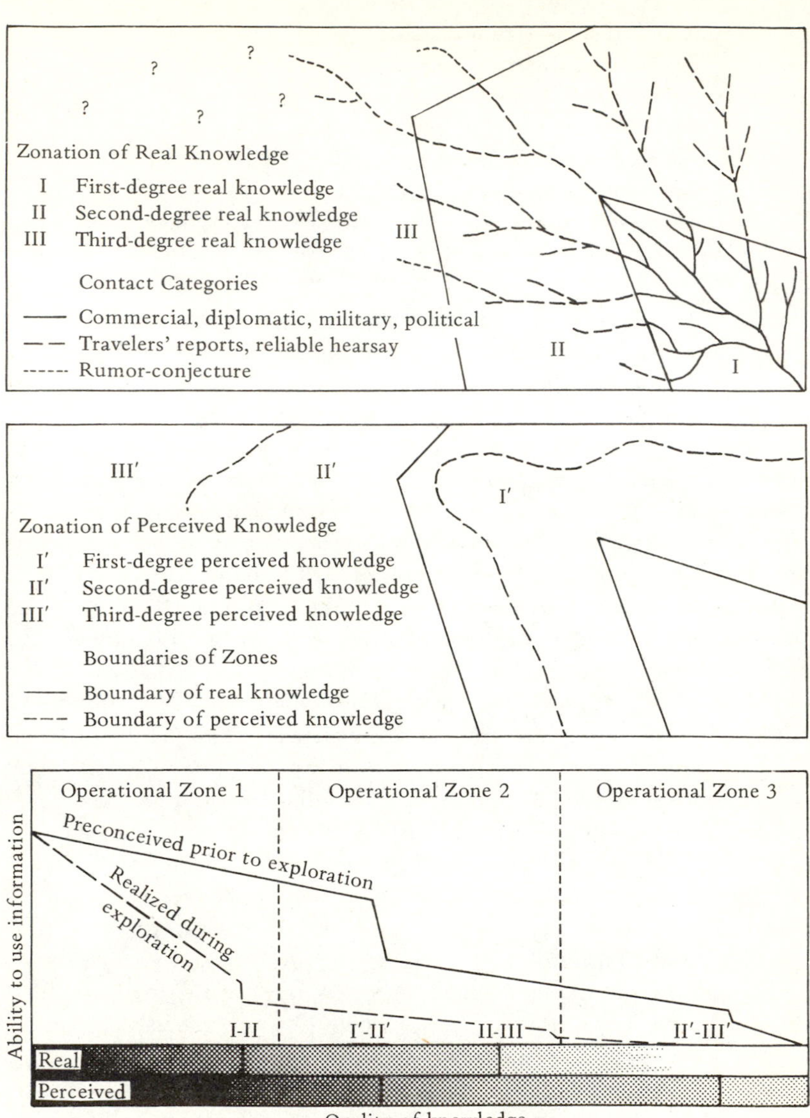

Fig. 9.2 Zonation of Real Knowledge, Zonation of Perceived Knowledge, and Perceived and Real Knowledge in Relation to Operational Zones. (John L. Allen, "An Analysis of the Exploratory Process: The Lewis and Clark Expedition of 1804-1806." Reprinted from the *Geographical Review*, Vol. 26, No. 1, January, 1972, copyrighted by the American Geographical Society of New York. Reprinted by permission.)

Lewis and Clark to shape a new image of the country they explored. In the course of developing this new, more realistic image of the West, many earlier misconceptions were shed including the idea of the Passage to India, so important as a motive for the mission. The explorers' original perception of their information as better than it actually was illustrates a common human tendency to decrease what is actually unknown by filling gaps in knowledge with what is believed to be known. It seems likely that this process operates at other scales as well, though perhaps to a different degree at different scales.

It may be that, as the total amount of information increases with larger portions of the environment, the proportion of first-degree knowledge held by the individual may diminish. Denis Wood has made some comparisons of sketch maps by the same individual at different scales.[8] He contends that people have similar numbers of image elements for any level of environmental ordering. For example, if a person puts twenty items on his sketch map of the city, he would probably also put twenty items on his map of the neighborhood and twenty items on his map of the world. This individual constancy at different scales proved true for the maps of block façades and barrios (neighborhoods) in San Cristobal, Mexico. Wood suggests that, regardless of the size and complexity of a given level, the number of orienting and navigational cues is relatively constant for a given individual. The amount of knowledge is related to degree of involvement, which might be expected to remain relatively constant at all the different scales of the environment that directly impinge on daily life. Because larger areas are more complex, the individual is familiar with a smaller portion of the significant features. Regions such as the room, the home, and the neighborhood are small enough to be perceived directly. At the other extreme are regions such as the nation, which are conceived as simply abstractions, so that images and ideas replace direct perceptions.

There is a great difference in the type, amount, and frequency of feedback from the environment experienced at these different scales that changes the quality of the knowledge. With less direct feedback there is likely to be a greater discrepancy between perceived and real knowledge because the individual's information will be based not on firsthand observation but on the observations or opinions of other people. In such a situation there is an increased probability of obtaining "second-" or "third-degree knowledge" (to use Allen's terms). This follows because what the individual learns from other observers may vary in quality from completely reliable observations to rumor, conjecture, or hearsay information. Futhermore, the individual will have less knowledge based on personal observation with which to assess the quality of the knowledge obtained. This helps to explain the dangerous degree of delusion seen in images at the international level. Here discrepancies between perceived and real knowledge become great enough to foster such unrealistic contradictions as the mirror-image phenomenon, described in Chapter 8, in

which two contending parties hold almost opposite views of the same situation.

Though gross misconceptions may be more common at the international level, the same lack of discrimination between perceived and real knowledge is a factor planners must contend with at other scales as well. Bizarre notions of ghetto areas or various ethnic groups may hamper planning efforts within a city, and building designs may be based on misconceptions from lack of real knowledge of the users' behavior. The literature offers example after example to show the unforeseen problems resulting from failure to consult the persons most concerned with a specific environmental change. When viewed from the perspective of different zones of knowledge, these problems may be seen as a direct result of planning based on something less than first-degree knowledge. The less familiar the planner is with the people-environment interactions in the system being altered, the more likely it is that essential features will be overlooked.

Most people's parochial viewpoints can be explained by the small-group origin of knowledge of the environment noted in the introductory chapter. As De Long pointed out, many of our concepts are acquired in an emotional manner in the context of a small group. As long as the information from the small group does not transcend the immediate area in which they live, there is little likelihood of developing the broader perspectives needed for more empathetic linking of various world areas into larger systems.

Professional perspectives on environmental perception

Parochial viewpoints are not limited to the public. Another type of narrowed perspective may be seen in each profession. In reviewing the research in this book, we have seen that each field has a unique characteristic focus. This shows up in the scale of the particular field and the type of phenomena studied.

At each succeeding scale we find different types of social scientists and design and planning professionals. As the scale changes, new groups are added and others fade out. Among the social scientists the smaller-sized areas are dominated by psychologists. The environment they are most concerned with is the social environment, consisting of other people. Sociologists commonly work at the scale of the neighborhood and the city where the built environment is a major concern. Geographers and anthropologists become more numerous at the regional scale where the natural environment assumes greater importance. At the world scale we find political scientists, whose focus is on a largely symbolic environment. Similarly, the design professionals shift with scale. Industrial designers, then architects, then urban and regional planners succeed each other as the scale is shifted from smaller to larger areas. There is a tendency for specific

groups working at a particular scale to remain isolated, unaware of similar work at their own or other scales.

Each profession is interested in only one aspect of the world. This is understandable in terms of the major orienting ideas and chief conceptual units of each field.[9] Table 9.1 shows an example of what some of these are for the social scientists.

It is interesting to compare the main organizing ideas and chief conceptual units of each social science with the major focus of various authors discussed in the preceding chapters. The psychologists, for example, examine the individual's reaction to the segment of the environment under consideration. Typically, they are concerned with personality differences and with the environment consisting of other people. Thus Sommer, at the scale of the room, studies the individual's reaction to spatial invasion; Barker and Wright, the way the small town settings impinge on the individual. Part of the fascination of Milgram's study of the city is based on this. Although the segment of the environment investigated is not the usual one for a psychologist, the focus on the individual remains. This provides a unique psychological perspective of the city, in striking contrast to the more usual sociological investigation focusing on social groups, as in the studies of Boston's West End. Here the contrasting environmental images of slum dwellers and urban renewal officials reflect social class differences. Repeated examples have also appeared of the anthropologist's concern with culture and the geographer's firm focus on particular landscapes. Because each of the social sciences is concerned with a different aspect of reality, it is easy to understand why they might have trouble communicating with each other, and an even more fundamental communication gap occurs between social scientists in general and those in the design and planning professions.

The visual semantic-communication gap

The literature on environmental perception behavior and planning is replete with references to the problem of communication between the social

Table 9.1 Chief Conceptual Unit

Field	Central organizing idea	Chief conceptual unit
History	Time	Event
Geography	Place	Located area
Political science	Power	State
Economics	Scarcity	Market
Sociology	Social behavior	Social system
Anthropology	Culture	Cultural system
Psychology	Personality	Individual

SOURCE: J. N. Rosenau, *International Politics and Foreign Policy*, Macmillan Publishing Co., Inc., New York, 1961, pp. 24-35. Reprinted by permission.

or behavioral scientists and various design and planning professionals.[10] While the social or behavioral scientists are justifiably cautious about implementing their first findings, those in the planning and design professions are anxious to incorporate the latest insights into a concrete design. Each tends to lose patience with the other.

This fundamental problem has been described by Edward Ostrander as "the visual-semantic communication gap."[11] He contends that "the designer-architect places considerable reliance upon visual modes of cognizing and communicating while the behavioral scientist turns to the semantic mode."[12]

In seeking information, the designer will turn to visual forms while the social scientist will use written sources. The two groups think in different ways, which makes communication difficult. The social or behavioral scientist generally engages in analytical thinking, separating things into their parts and looking for the relationship of these parts to each other. This contrasts with the designer, whose usual mode of thought is synthetic. The designer combines or synthesizes various elements into a single or unified whole. Further differences in usual mode of thought may be seen in the contrast between the scientific method of information-processing, which takes one thing at a time, and the method of the architect or designer which considers simultaneously a whole host of dimensions, mainly in visual form. Furthermore, the level of abstractions is different. The social scientist in general deals with abstract concepts or theoretical models, while the designer most often works on a rather concrete level, with sketches of specific rooms, buildings, or sites. Interestingly, these contrasting modes of thought may result from genetically rooted individual differences, for Ostrander notes that:

> recent work on brain hemisphere dominance suggests that semantic, analytic, and sequential information processing are primarily left hemisphere functions while pre-verbal, synthesizing, and simultaneous information processing activities are primarily right hemisphere functions.[13]

Thus we might expect that the original occupational choice is based on individual differences in hemisphere dominance.

A simple illustration of the different research directions that occur as a result of professional stances is provided by tracing the fate of the image. As first sketched out by Boulding, a social scientist, the image was an abstract theoretical concept. When Lynch took up the concept, the planner's concern for concrete practical application of current knowledge was evident. He pioneered a method of investigating public images of the city to use this information to design better cities. Some of Lynch's ideas were carried further by Downs and Stea, both social scientists, and by Appleyard, a planner. The aspects of Lynch's work that intrigued Downs and Stea were the basic scientific questions of what these images are and how they are

formed. Their investigations were focused not on the concrete, practical planning problems but on the more abstract, theoretical questions involved in cognitive maps. Appleyard, on the other hand, took Lynch's technique one step further along the road to practical planning applications. He set up scales for evaluating why certain buildings are known and appear in cognitive maps. These scales provide planners with concrete guidelines to apply in designing each building to achieve the degree of prominence it deserves. Authors—such as those cited in this book who are all well known for advancing interdisciplinary research—reflect professional perspectives.

The aim of creating an environment in harmony with human needs and goals has brought together professionals with widely differing perspectives. For the designer and the social scientist a long period of familiarization with each other's methods and aims seems necessary. As Kenneth Craik has pointed out:

> . . . the magnitude of the methodological and empirical groundwork that must be established as a basis for a mature branch of research makes it imperative to think in terms of decades rather than in months or years, and makes it incumbent upon behavioral scientists in this field to be humble in their advice and proclamations as well as incumbent upon environmental planners and designers to be patient in their expectations.[14]

Reality worlds in collision

The preceding discussion on the parochial viewpoints held by people, on the varying perspectives of the social sciences, and on the differing approaches of designers and social scientists illustrates that there are many ways of constructing reality. As the opening quotation in this chapter indicates, these differing models of reality become real in their consequences as people act on them. As long as the model bears some relationship to reality, one can serve as well as another. In the past people in different areas lived out their lives with minimum conflict because the areas remained isolated from each other. But today, with rapidly growing populations causing increased pressure on world resources and greater mobility resulting in increased contact among varying groups, reality worlds are colliding.

> In a world that is becoming smaller and smaller each of us will sooner or later have to deal with a "foreign soil," or in terms of our model, another way of constructing reality. The Spaceship Earth concept and the Ecology movement are drawing attention to the fact that we not only will have to learn to live with whom we formerly considered foreigners, but will have to learn to cooperate with them.[15]

To do so we must accept the relativity of our own reality worlds. We must remain open and respectful toward other realities, no matter how strange

or unbelievable they may seem to us. This is easier said than done, for, as was pointed out in the introduction, much of our behavior is totally outside-awareness and irrational. Tibbetts and Esser suggest that what is required is "a transcendence of biological and cultural limitations through *empathy*: the *intellectual* identification with the needs, feelings and thoughts of others in order in promote their welfare."[16] With this type of attitude and the kind of information that can be acquired through a theory of people-environment relations that bridges the gap between arts and sciences, we may be able to build a better environment on spaceship earth.

Notes

1. Roger G. Barker, *Ecological Psychology: Concepts and Methods For Studying the Environment of Human Behavior,* Stanford University Press, Stanford, Calif., 1968, p. 154.
2. Raymond G. Studer,"Dynamics of Behavior—Contingent Physical Systems," in Harold M. Proshansky, William H. Ittelson, and Leanne G. Rivlin, *Environmental Psychology: Man and His Setting,* Holt, Rinehart and Winston, Inc., New York, 1970, pp. 56-76.
3. *Ibid.,* p. 72.
4. Denis Wood, *Fleeting Glimpses,* Clark University Cartographic Laboratory, Worcester, Mass., 1971, p. 199.
5. Robert Sommer, "Space Time on Prairie Highways," *AIP Journal* (July, 1967), 274-276.
6. John L. Allen, "An Analysis of the Exploratory Process: The Lewis and Clark Expedition of 1804-1806," *Geographical Review,* 62, No. 1 (January, 1972), 13-39.
7. *Ibid.,* p. 14.
8. Denis Wood, *op. cit.,*p. 215.
9. Charles A. McClelland, "The Social Sciences, History, and International Relations," in James N. Rosenau, ed., *International Politics and Foreign Policy,* Glencoe Press, Inc., New York, 1961, pp. 24-35.
10. See, for example, Lawrence Wheeler, "The Designer and the Behavioral Scientist: A Difficult, Exciting, Cooperative Adventure," based on a paper given at the University of Wisconsin, October, 1972, to open a conference entitled "Designing Architectural Interiors to Support Human Performance."
11. Edward Ostrander, "The Visual-Semantic Communication Gap: A Model and Some Implications For Collaboration Between Architects and Behavioral Scientists," *Man-Environment Systems,* 4, No. 1 (January, 1974), 47-53.
12. *Ibid.,* p. 48.
13. *Ibid.,* p. 47.
14. Kenneth Craik, "The Comprehension of the Everyday Physical Environment," *Journal of the American Institute of Planners,* 34, No. 1 (January, 1968); also reprinted in Proshansky *et al., Environmental Psychology,* pp. 646-658, quote p. 657.
15. Paul Tibbetts and Aristide Esser, "Transactional Structures in Man-Environment Relations," *Man-Environment Systems,* 3, No. 6 (November, 1973), 441-468, quote on p. 463.
16. *Ibid.,* p. 465.

Author index

Subject index

BCDEFGHIJ–A–79876